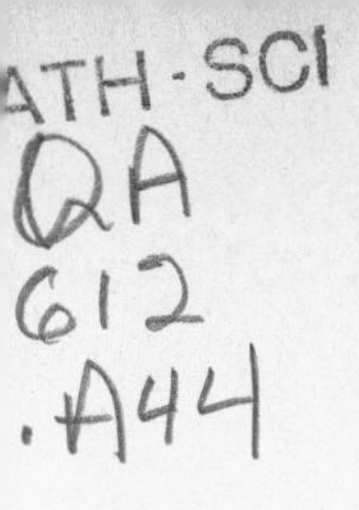

FOREWORD

From July 25 to August 12,1977, the Canadian Mathematical Society sponsored a Workshop and Conference in Algebraic Topology at the University of British Columbia, in Vancouver.

The Workshop, which involved the active participation of Graduate Students, consisted of a series of informal lectures devoted to the presentation of new theories and of background material related to talks to be delivered later on; furthermore, during this part of the meeting, there were discussions about the research work being conducted by Graduate Students. The speakers of the Workshop were: E.Campbell*, A.Dold, R.Douglas, S.Feder, P.Heath, P.Hoffman, R.Kane, J. McCleary*, C.Morgan*, L.Renner* and C.Watkiss.

The lecturers of the Conference proper were: R.Body, P.Booth, A.Dold, R.Douglas, S.Feder, H.Glover, P.Heath, P.Hilton, R.Kane, A.Liulevicius, G.Mislin, S.Segal, L.Siebenmann, F.Sigrist, D.Sjerve, V.Snaith, J.Stasheff, U.Suter, M. Tangora and C.Watkiss.

The articles printed in these Proceedings are based on talks given during the Conference; we observe that not all the talks are represented here since some speakers elected not to submit a paper.

The published papers have been divided into four areas:

A. Rational Homotopy Theory;

B. Cohomology Theories; Bundle Theory;

C. Homotopy Theory; Nilpotent Spaces; Localization;

D. Group Cohomology; Actions.

The general index has been prepared so as to make this division clear. In each area the articles appear in alphabetical order by the name of the author or first author.

The list of addresses of all contributors and the names of all participants are given at the end of this volume.

P. Hoffman

R. Piccinini

D. Sjerve

*Graduate Students

CONTENTS

THE UNIQUENESS OF COPRODUCT DECOMPOSITIONS FOR ALGEBRAS OVER A FIELD[†]

by

Roy Douglas[*]

1. Introduction

Coproduct decompositions of various types of graded F-algebras will be considered, where F is an arbitrary fixed field. The main result will be a "unique factorization" for such decompositions. For some of these types of algebras, the coproduct is just the appropriate type of tensor product.

We will consider various categories of F-algebras and their associated commutative semigroups of isomorphism classes of objects (where the binary semigroup operation is induced by coproduct). The "unique factorization" results will then be expressed by the statements that the above semigroups are <u>free</u> commutative semigroups.

Examples of suitable categories of F-algebras are the following:

(1) the category of all connected, finitely generated, associative, <u>commutative</u>, (graded) F-algebras.

(2) the category of all connected, finitely generated, associative, <u>graded-commutative</u>, (graded) F-algebras.

(3) the category of all connected, finitely generated, (graded) Lie algebras over F.

The discussion below proves the "unique factorization" assertion in examples (1) and (2), (1) being a corollary of (2). Moreover, this discussion may be generalized to give "unique factorization" results for many other types of graded algebras, including example (3).

---

* Research partly supported by the National Research Council of Canada.
† The content of this address is the result of joint work with R. A. Body.

In case F is a perfect field, this result is demonstrated in [2] .

In case F has characteristic zero, certain (large and interesting) classes of F-homotopy types of topological spaces satisfy a unique decomposition property with respect to direct product. This is demonstrated for two such classes of F-homotopy types in [2] and [3] , respectively.

All these results, somewhat reminiscent of the Krull-Schmidt theorem, are proved by a study of the conjugacy properties of certain linear algebraic groups of automorphisms.

2. Unique Factorization for Coproducts

Let F be an arbitrary field.

$\Omega(F)$ (resp., $\Delta(F)$) will denote the category whose objects are all associative, graded-commutative (resp., strictly commutative), connected, finitely generated F-algebras, and whose morphisms are all degree preserving F-algebra homomorphisms. For brevity, the objects of $\Omega(F)$ will be referred to as F-algebras. The graded-commutative (resp., commutative) tensor product $\otimes$ is the coproduct in $\Omega(F)$ (resp., $\Delta(F)$).

The isomorphism classes of objects of $\Omega(F)$ (resp., $\Delta(F)$) form a commutative semigroup $\omega(F)$ (resp., $\delta(F)$), where the binary operation is induced by $\otimes$ (the unit is the "zero object" F). The unique factorization of F-algebras (with respect to $\otimes$) is expressed by:

<u>Theorem 1</u>. $\omega(F)$ is a <u>free</u> commutative semigroup.

By doubling the gradation degrees we obtain the following

<u>Corollary</u>. $\delta(F)$ is a <u>free</u> commutative semigroup.

Before proving Theorem 1, several useful observations will be recorded. Let A be an F-algebra. A is <u>non-trivial</u> if $A \neq F$. We say a non-trivial A is <u>irreducible</u> if A is not the tensor product of two non-trivial F-algebras.

<u>Definition</u>. A finite set $\{e_1, \ldots, e_n\}$ of (graded) F-algebra endomorphisms of A, will be called a <u>splitting</u> (of A) if the following conditions are satisfied:

(1) $e_i \cdot e_i = e_i$, $i = 1, \ldots, n$ (Idempotent)

(2) $e_i \cdot e_j = 0$, $i \neq j$ (Orthogonal).

(0 is the trivial endomorphism, which factors through the zero object.)

(3) Each F-algebra, Image (e_k), is irreducible, $k = 1, \ldots, n$

(4) The canonical morphism

$$\bigotimes_{k=1}^{n} \text{Image}(e_k) \to A$$

is an isomorphism in $\Omega(F)$.

Two splittings of A, $\{e_1, \ldots, e_n\}$ and $\{f_1, \ldots, f_n\}$ are said to be <u>equivalent</u> if, for some permutation σ, the F-algebras Image(e_i) and Image$(f_{\sigma(i)})$ are isomorphic, $i=1, \ldots, n$.

If α is an automorphism of A, then $\{e_1, \ldots, e_n\}$ and $\{\alpha.e_1.\alpha^{-1}, \ldots, \alpha.e_n.\alpha^{-1}\}$ are equivalent splittings.

<u>Proposition 2</u>. If $\{e_1, \ldots, e_n\}$ and $\{f_1, \ldots, f_m\}$ are splittings of an F-algebra A and $e_i.f_j = f_j.e_i$, for all $i = 1, \ldots, n$ and $j = 1, \ldots, m$, then these splittings are equivalent. (Thus, $n = m$.)

(The proof of this proposition is elementary; for details see [2] , Lemma 2.)

Theorem 1 follows easily from Proposition 2 and the following lemma.

<u>Lemma 3</u>. If $\{e_1, \ldots, e_n\}$ and $\{f_1, \ldots, f_m\}$ are splittings of an F-algebra A, then there exists an automorphism $\alpha : A \to A$ such that $\alpha.e_i.\alpha^{-1}.f_j = f_j.\alpha.e_i.\alpha^{-1}$ for all $i = 1, \ldots, n$ and $j = 1, \ldots, m$.

3. Proof of Lemma 3.

Let End(A) be the semigroup of all endomorphisms of F-algebra A, and let Aut(A) be the group of invertible elements of End(A).

Since $A = \bigoplus_{i \geqslant 0} A^i$ is finitely generated as an F-algebra, there is an ℓ such that A is generated (as an F-algebra) by the finite dimensional F-vector subspace $V = \bigoplus_{i=1}^{\ell} A^i$.

The restriction, $\mathrm{End}(A) \to \mathrm{Hom}_F(V,V)$, and its restriction $\mathrm{Aut}(A) \to GL_F(V)$ are injective, representing End(A) and Aut(A) as sets of square matrices with coefficients in F.

End(A) is the set of zeros of an obvious set of linear and quadratic polynomials (in the entries of the matrices) with coefficients in F.

Let K be an algebraic closure of F and let $W = V \otimes_F K$. The above set of polynomials (with coefficients in F) then defines a variety E in the K-affine space $\mathrm{Hom}_K(W,W)$. Moreover, E is a closed set in the Zariski F-topology on $\mathrm{Hom}_K(W,W)$. (See [4] for definitions.) Of course, End(A) is the set of F-rational points of E.

Similarly, Aut(A) is the set of F-rational points in the affine algebraic group $G = E \cap GL_K(W)$.

There is a finite, purely inseparable extension field L of F ($F \subset L \subset K$), such that E and G are defined over L. (In case F is perfect, L = F.)

Now consider the splitting

$\{e_1, \text{---}, e_n\} \subset \mathrm{End}(A)$ for F-algebra A.

Let $A_k = \mathrm{Image}(e_k)$, a sub-F-algebra of A, with $\bigotimes_{k=1}^{n} A_k = A$.

$$A^p = \bigoplus_{\sum_{k=1}^{n} p_k = p} \bigotimes_{k=1}^{n} A_k^{p_k} \quad \text{and} \quad V = \bigoplus_{\sum_{k=1}^{n} p_k \leq \ell} \bigotimes_{k=1}^{n} A_k^{p_k}.$$

Thus, $W = \bigoplus_{\sum_{i=1}^{n} p_i \leq \ell} \bigotimes_{i=1}^{n} (A_i^{p_i} \otimes_F K)$, where the n-fold tensor product is constructed over K.

For each $i=1,2,\ldots,n$ and each $t \in F^*$, the automorphism $\lambda_i'(t) \in \mathrm{Aut}(A) \subset GL_F(V)$ is defined to be scalar multiplication by t^s on a direct summand $\bigotimes_{j=1}^{n} A_j^{p_j}$ of V, where $s = \sum_{j \neq i} p_j$.

$\lambda_i : K^* \to G$ is a one parameter subgroup of G, where λ_i is defined similarly in terms of such an eigenspace decomposition of W.

Let S be the subgroup of G generated by $\{\lambda_i(t) | t \in K^*, i = 1, \ldots, n\}$.
Then S is an L-split torus of G. (cf. [4], p.200.)

Observe that $\{e_1, \ldots, e_n\} \subset \overline{S} \subset E$, where $\overline{S}$ is the K-Zariski closure of S in E.

In fact,

$$e_i \in \overline{\lambda_i(K^*)} \subset E, \text{ since K is an infinite field.}$$

Thus, there is a maximal L-split torus $T_e \subset G$, such that $\{e_1, \ldots, e_n\} \subset \overline{T}_e \subset E$.
Similarly, there is a maximal L-split torus $T_f \subset G$, such that $\{f_1, \ldots, f_n\} \subset \overline{T}_f \subset E$.

Using an unpublished result of Borel and Tits [5] it follows that there is an L-rational point $\beta \in G(L)$ such that $\beta \cdot T_e \cdot \beta^{-1} = T_f$. (*)

In case F is a perfect field, we have L = F, and $\beta \in G(F) = \text{Aut}(A)$ is the required automorphism α of Lemma 3. This follows from the fact that the closure of a torus is a commutative set of endomorphisms, and the fact that conjugation by β is a homeomorphism.

In case F is not perfect, a finite iteration of the Frobenius morphism ϕ takes L into F: $\phi^s(L) \subset F$.

Let $\alpha = \phi^s(\beta) \in G(F) = \text{Aut}(A)$ and observe that $\phi^s(e_i) = e_i \in \text{End}(A) \subset E$.
Notice $\phi^s(T_e)$ is dense in T_e, since F is infinite. (Recall that finite fields are perfect.)

Thus, $\{e_1, \ldots, e_n\} \subset \overline{\phi^s(T_e)}$. Similarly, $\{f_1, \ldots, f_m\} \subset \overline{\phi^s(T_f)}$. Again, $\overline{\phi^s(T_e)}$ and $\overline{\phi^s(T_f)}$ are each commutative sets of endomorphisms.
Applying ϕ^s to (*) we obtain

$$\alpha \cdot \overline{\phi^s(T_e)} \cdot \alpha^{-1} = \overline{\phi^s(T_f)}$$ which implies the conclusion of Lemma 3.

Q.E.D.

References

1. R. Body and R. Douglas, Rational Homotopy and Unique Factorization, Pacific J. Math. (to appear).

2. R. Body and R. Douglas, Tensor Products of Graded Algebras and Unique Factorization, Amer. J. Math. (to appear).

3. R. Body and R. Douglas, Unique Factorization of Rational Homotopy Types Having Positive Weights, Comment. Math. Helv. (submitted).

4. A. Borel, Linear Algebraic Groups, W. A. Benjamin, New York, 1969.

5. Private letter from A. Borel to R. Douglas, stating the following unpublished result of A. Borel and J. Tits:
"If G is a connected affine algebraic group over a field k, then its maximal k-split tori are conjugate under G(k)."

Rational homotopy-obstruction and perturbation theory*

James Stasheff

Just how algebraic is algebraic topology? Traditionally the emphasis has been the study of functors turning topology into algebra. For example, for any commutative ring R, the cohomology groups $H^*(X;R)$ of a space X form a graded commutative algebra over R.

One can turn the process around and ask:

1) Given a graded commutative algebra H over R, is there a space X such that $H^*(X;R) \approx H$?

2) Given an abstract (purely algebraic) isomorphism $\phi: H^*(Y) \approx H^*(X)$, is there a map $f: X \to Y$ inducing ϕ?

3) Classify all homotopy types X such that $H^*(X;R) \approx H$.

In this series of lectures I will restrict myself to R = Q = the rationals, since the situation in that case is under much better control.

The positive answer to question 1) has been known for some time, at least since the work of Quillen [9], but our method of answering will be quite different. First of all, it is strongly motivated by topology and is comparatively naive algebraically. Secondly, it leads to an obstruction theory for 2) in the case of isomorphism which although

---

* These notes are a mild revision of the lectures given at the Conference. The first three lectures were based on my paper with Stephen Halperin (Toronto), Obstructions to Homotopy Equivalences to appear in Advances in Mathematics, hopefully in 1982! That paper contains complete details, computations, etc. - these lectures hope to present the main concepts and a clear indication of techniques. The final lecture concerns work in progress with Mike Schlessinger (University of North Carolina) although the form of presentation and even some of the results have been influenced by activity at this Conference.

purely algebraic has essentially the form familiar in topology. Finally it opens a new approach to the classification of rational homotopy types which brings out certain relations with algebraic geometry.

Our emphasis throughout is on doing algebra but in a way which is guided by topology.

To make the transition from topology to algebra,we need two key ideas: For simplicity of exposition I will restrict myself to simply connected spaces of the homotopy type of a CW complex of finite type, henceforth called just spaces. Extensions of the theory exist but at the expense of being awfully precise about matters such as the action of the fundamental group.

DEFINITION. A rational homotopy equivalence $f:X \to Y$ is a map of spaces such that

$f^*:H^*(Y,Q) \to H^*(X;Q)$ is an isomorphism or equivalently

$f_*:\pi_*(X) \otimes Q \to \pi_*(X) \otimes Q$ is an isomorhpism.

Two spaces X,Y are rationally equivalent if there exists maps $X = X_0 \to X_1 \leftarrow X_2 \to X_3 \leftarrow \dots \to X_n = Y$ which are rational homotopy equivalences.

From this point of view, it is helpful to consider rational spaces i.e. spaces such that $H^*(X;Z)$ or equivalently $\pi_*(X)$ are vector spaces over the rationals, finite dimensional in each degree. For such spaces a rational homotopy equivalence is an ordinary homotopy equivalence. A most improtant example is $K(Q,n)$, the Eilenberg-MacLane space of type (Q,n), i.e. $\pi_i = 0$ except $\pi_n \simeq Q$. By induction it is easy to prove

$H^*(K(Q,2n)) \simeq Q[x_{2n}]$, the polynomial algebra on a single generator of dim 2n while

$H^*(K(Q,2n-1)) \simeq E(x_{2n-1})$, the exterior algebra on one generator of dim 2n-1.

Notice that $K(Q,2n-1)$ is rationally equivalent to S^{2n-1} while $K(Q,2n)$ is rationally equivalent to ΩS^{2n+1}.

We see that for rational coefficients any free graded commutative algebra can be realized, e.g., as a product of odd dimensional spheres and loop spaces of even dimensional spheres or as a product of $K(Q,n)$'s.

As soon as relations appear in the algebra, the situation is more complicated. For example $Q[x_{2n}]/x^2 = 0$ can be realized as $H^*(S^{2n})$ because we happen to recognize the algebra of a familiar space. Alternatively we can build S^{2n} rationally without prior knowledge as follows:

Consider $A = Q[x_{2n}] \otimes E(x_{4n-1})$ with differential d acting as a derivation such that $dx_{4n-1} = (x_{2n})^2$. It is easy to compute that $H(A,d) \approx Q[x_{2n}]/x_{2n}^2 = 0$. Now fibre spaces appear in cohomology as such twisted tensor products so we reinterpret A in terms of a fibration

$$\begin{array}{ccc} K(Q,4n-1) & = & K(Q,4n-1) \\ \downarrow & & \downarrow \\ X & \to & \mathcal{L}K(Q,4n) \\ \downarrow & & \downarrow \\ K(Q,2n) & \xrightarrow{x^2} & K(Q,4n). \end{array}$$

Our computation shows X is rationally equivalent to S^{2n}. In some sense, A is a rational cochain algebra for S^{2n}. For more complicated algebras, we need to look more carefully at the cochain level. Commutative cochain algebras are the key to this and other problems we attack.

The original idea comes from differential topology - de Rham's theorem involving differential forms on a manifold. It is not difficult to adapt the idea to simplicial complexes or simplicial sets, e.g. the singular complex of a space X.

DEFINITION: The algebra $A^*(\Delta^n)$ of rational (polynomial) forms on the standard n-simplex Δ^n is the graded commutative algebra

$$Q[t_1,\ldots,t_n] \otimes E(dt_1,\ldots,dt_n)$$

with $d{:}t_i \to dt_i$ extended as a derivation.

Thus a rational form of degree q looks like a sum of terms

$$p(t_1,\ldots,t_n)dt_{i_1}\wedge \ldots \wedge dt_{i_q}$$

where p is a polynomial with rational coefficients.

DEFINITION. For a simplicial set $S = \{S_n\}$, _the algebra_ $A^*(S)$ _of rational forms_ on S consists of collections $\{w_\alpha | \alpha \in S\}$ where w_α is a rational form on the simplex α of S such that $\partial_i^* w_\alpha = w_{d_i\alpha}$ where $\partial_i : \Delta^{n-1} \to \Delta^n$ is the i-th face. That is the w_α are compatible in that the restriction to a face is again in the collection.

For our work, all that matters is that $A^*(S)$ satisfies the following definition and theorem.

DEFINITION: A _c.d.g.a. (commutative differential graded algebra)_ A is a graded ($A = \bigoplus_{n \geq 0} A^n$) vector space with a multiplication

$$\wedge : A^p \otimes A^q \to A^{p+q}$$

making it a commutative graded algebra i.e. $\wedge$ is associative and

$$w^p \wedge w^q = (-1)^{pq} w^q \wedge w^p$$

with a differential

$$d : A^p \to A^{p+1}$$

such that $d^2 = 0$ which is a derivation

$$d(w^p \wedge w^q) = dw^p \wedge w^q + (-1)^p w^p \wedge dw^q$$

Theorem (Simplicial de Rham [4], [2], [11]):
$H(A^*(S,d)) \approx H^*(S;Q)$, the rational cohomology as a simplicial set.

Rational forms provide a functor from simplicial sets to c.g.d. a's and in turn, via the singular complex S, from topological spaces: Top spaces $\to$ Δ-sets $\to$ C.g.d.a.'s/Q.

The main result of Quillen (now done with improved clarity and efficiency by several others [2], [3], [4]) is the following:

THEOREM: There is a suitably algebraic notion of homotopy for c.g.d.a 's such that the homotopy category of simply connected c.g.d.a's over Q is equivalent to the rational homotopy category of simply connected simplicial sets or equivalently to the homotopy category of rational spaces.

Our program will establish this result by working in the opposite direction.

We break the process into two steps.

THEOREM 1. Given a connected c.g.a. H, there exists a rational space X_H such that $H^*(X_H) \simeq H$.

The space X_H constructed in the proof is called the formal space associated to H since it is completely determined, up to homotopy type, by H.

Theorem 1 is but the topological expression of a theorem in commutative graded algebra:

THEOREM 1' [cf. 8]: Given a connected c.g.a H over Q, there exists a c.g.d.a. A which is free as a c.g.a and a homology equivalence $(A,d) \xrightarrow{\rho} (H,0)$.

Since A is free as c.g.a., it is of the form $E(Z^{odd}) \otimes Q[Z^{even}]$ for some graded rational vector space Z. We will write simply ΛZ to denote the free c.g.a. on Z. We will construct ΛZ inductively so it has the form $\Lambda Z_0 \otimes \Lambda Z_1 \otimes \Lambda Z_2 \otimes \ldots$ where the lower index is just for the induction (it is like a resolution degree) and $d|Z_{n+1}: Z_{n+1} \to \Lambda Z_0 \otimes \ldots \otimes \Lambda Z_n$. Thus X_H can be built from successive fibrations of products of $K(Q,n)$'s by copying this differential. (See section below).

Our second major step in equating our two categories, topology and algebra, is:

THEOREM 2. Given a simply connected space X and the formal c.g.a. $(\Lambda Z,d)$ associated to $H(X)$, there is a perturbed differential D on ΛZ (i.e. $D-d: Z_{n+1} \to \Lambda Z_0 \otimes \dots \otimes \Lambda Z_{n-1}$) and a homology equivalence $(\Lambda Z,D) \xrightarrow{\pi} A^*(X)$.

Moreover, there is a very strong uniqueness theorem - but that we will take up in some detail later.

It is these models $(\Lambda Z,D)$ for a space with cohomology $H^*(X) \simeq H$ that allow us to construct an obstruction theory (in the usual topological sense) for realizing an isomorphism $\phi: H^*(Y) \simeq H^*(X)$ by an equivalence $X \to Y$.

On the other hand, every perturbation D such that $D^2 = 0$ can be realized by a space. In fact, the set of homotopy types of rational spaces realizing a given H is (over) determined by the space of perturbations, so we are lead to a problem familiar to algebraic geometers, that of describing a space of moduli. Some curious relations with known results in the deformation theory of pure algebra have been discovered by Mike Schlessinger and will be the subject of our final lecture.

2. The Tate-Jozefiak resolution and its realization.

But enough of the general setting - let us get down to the specifics of realizing a given c.g.a. H by a free cgda ΛZ. Halperin and I thought we were developing something new for topology, but the purely algebraic version had indeed been done in varying degrees by algebraists, most completely and closely to our approach by Jozefiak [8]. As indicated by his title, when H itself is ungraded (or concentrated in degree 0), the construction reduces to that of Tate, who was influenced by John Moore and the method of killing cohomology by fibrations!

We will parallel our construction with the illustrative example: $H = H^*(S^2 \vee S^2)$.

For any connected graded algebra A, we let $A^+ = \bigoplus_{n \geq 1} A^n$ and QA = the module of indecomposables = $A^+/A^+ \cdot A^+$ so that a basis for QA is a generating set for A.

Now given a connected c.g.a. A, let $Z_0 = QH$ and let ΛZ_0 be the free commutative algebra on Z_0, i.e. $E(Z_0^{odd}) \otimes Q[Z_0^{even}]$. Any monomial in ΛZ_0 will be said to have bottom degree 0 and top degree the sum of the gradings of the indecomposable factors.

Choose (additively) a section $Z_0 \to H$, i.e. representative of the indecomposable classes and extend to a multiplicative map $\rho : \Lambda Z_0 \to H \to 0$. The kernel is a ΛZ_0-module . Let Y_1 be a minimal vector space generating the kernel as ΛZ_0-module. Let $Z_1 = s^{-1} Y_1$ i.e. an isomorphic vector space with grading shifted down by 1 so that $Z_1 \underset{s}{\overset{\simeq}{\to}} Y_1$ is a map of degree 1.

Consider $\Lambda Z_0 \otimes \Lambda Z_1$ with differential d defined by the isomorphism

$$Z_1 \simeq Y_1 \to \Lambda Z_0$$

and extended as a derivation with respect to the top degree. A monomial in $\Lambda Z_0 \otimes \Lambda Z_1$ is given a lower degree by summing the lower degrees of the factors, each element of Z_1 having lower degree 1, so, for example, writing $x^i \in Z_0^i$ and $y^j \in Z_1^j$ to indicate degrees, $x^p x^q y^r y^s$ has top degree p+q+r+s and bottom deg 2. With these gradations, d has top degree 1 and bottom degree -1. Thus homology can be defined as usual and is again bigraded. Extend ρ to be zero on Z_1 and then multiplicatively so that $\rho^*: H_0(\Lambda Z_0 \otimes \Lambda Z_1) \approx H$ but H_i may be non-zero for $i > 0$.

Example. Consider $H(S^2 \vee S^2) \approx Q[x,y]/x^2 = xy = y^2 = 0$. Take

$$\Lambda Z_0 = Q[x,y]$$

$$\Lambda Z_1 = E(u,v,w) \text{ with } du = x^2$$
$$dv = xy$$
$$dw = y^2.$$

Notice $xv-yu$ is a cycle but not a boundary. Indeed H_1 is generated by $xv-yu$ and $xw-yv$ so we can kill H_1 with

$$\Lambda Z_2 = Q[r,s] \text{ with } dr = xv-yu$$
$$ds = xw-yv.$$

Next we find $H(\Lambda Z_0 \otimes \Lambda Z_1 \otimes \Lambda Z_2)$ has $H_0 \approx H$

$$H_1 \approx 0$$

but H_2 is non-zero so we go on.

It's time for induction. Let

$$\Lambda Z_{(n)} = \Lambda Z_0 \otimes \ldots \otimes \Lambda Z_n.$$

Suppose $(\Lambda Z_{(n)}, d)$ has been constructed for some $n \geq 1$, so that d is homogeneous of bottom degree -1. Define Z_{n+1} by

$$Z^p_{n+1} = \left(H_n(\Lambda Z_{(n)}, d) \Big/ H_n(\Lambda Z_{(n)}, d) \cdot H^+_0(\Lambda Z_{(n)}, d) \right)^{p+1}$$

and extend d so that $d: Z_{n+1} \to (\Lambda Z_{(n)})_n \cap \ker d$ splits the projection of $(\Lambda Z_{(n)})_n \cap \ker d$ onto Z_{n+1}. Extend ρ to be zero on Z_{n+1}.

Now let $(\Lambda Z, d) \xrightarrow{\rho} (H,0)$ be the homomorphism of c.g.d.a.'s constructed in this way, with $Z = \sum_{n=0}^{\infty} Z_n$. It satisfies the requirements of Theorem 1'. When we build the corresponding space X_H, we will see $\bigoplus_{n \geq 0} Z^p_n \approx \pi_p(X_H)$. For $H = H(S^2 \vee S^2)$, X_H is rationally equivalent to $S^2 \vee S^2$. It is well known [7] that $\pi_*(S^2 \vee S^2) \otimes Q$ is the free Whitehead algebra on two generators (equivalently $\pi_*(\Omega(S^2 \vee S^2)) \otimes Q$ is a free graded Lie algebra on two generators). Thus Z_n is non-zero for all n and dim Z_n increases without bound as a function of n.

Several comments are in order. Just as in ordinary homological algebra, one can easily prove $(\Lambda Z,d)$ is uniquely determined up to homotopy. However, our particular construction chose a minimal vector space Z_{n+1} to kill $H_n(\Lambda Z_0 \otimes \ldots \otimes \Lambda Z_n)$. A simpler way of expressing the minimality is that $dZ \subset \Lambda Z^+ \cdot \Lambda Z^+$. If $(\Lambda Z',d') \underset{\rho'}{\to} H$ is any other minimal free c.g.d.a with ρ^* an isomorphism, then $(\Lambda Z',d')$ is <u>isomorphic</u> to $(\Lambda Z,d)$. We refer to $(\Lambda Z,d)$ as the bigraded or minimal model for H.

The construction of $(\Lambda Z,d)$ allows us to realize any connected c.g.a H as the cohomology of a rational space, namely by successive fibrations $E_{n+1} \to E_n$ corresponding to $\Lambda Z_{(n)} \to \Lambda Z_{(n+1)}$.

First E_0 realizes ΛZ_0 as a product of $K(Q,p)$'s. By induction assume E_n exists and a map

$$\Lambda Z_{(n)} \to A^*(E_n)$$

inducing an isomorphism in cohomology. Recall that dZ_{n+1} is identified with a quotient of $H_n(\Lambda Z_{(n)})$. Let $dZ_{n+1} \to H_n(\Lambda Z_{(n)})$ choose representatives. The free algebra ΛdZ_{n+1} can be realized by a product B_n of $K(Q,p)$'s and the map by a map $E_n \to B_n$. Let E_{n+1} be the induced fibration.

The good old reliable Koszul-Hirsch theorem shows $H^*(E_{n+1}) \approx H(\Lambda Z_{(n+1)})$ while the freeness of $\Lambda Z_{(n+1)}$ and the commutativity of $A^*(E_{n+1})$ allows this to be realized by a map $\Lambda Z_{(n+1)} \to A^*(E_{n+1})$.

Since we are dealing with simply connected spaces, the connectivity of Z_n is at least n+1 so the tower of spaces $\to E_{n+1} \to E_n \to \ldots$ determines a single space X_H with $H^*(X_H) \approx H$. We call X_H the <u>formal space</u> determined by H.

Notice that $\pi_*(X_H) \approx \oplus Z_n$ or in more detail

$$\pi_p(X_H) \approx \bigoplus_n Z_n^p$$

since we put in a $K(Q,p)$ for each basis element of Z_n^p.

We call any space Y <u>formal</u> if it is of the homotopy type of some X_H, necessarily $X_{H(Y)}$. Equivalently if $(\Lambda Z,d)$ is the bigraded model for $H(Y)$, then Y is formal iff there exists a homotopy equivalence $(\Lambda Z,d) \to A^*(Y)$.

Examples of formal spaces abound:

- products of $K(Q,n)$'s
- products of spheres
- wedges of spheres
- certain homogeneous spaces
- compact Kähler manifolds [3].

In particular G/H is formal if G and H are Lie groups of equal rank or if G/H is a symmetric Riemannian manifold. Every homogeneous space has cohomology of the form $H(E(P^{odd}) \otimes Q[P^{even}],d)$ with $d:P^{odd} \to Q[P^{even}]$ which looks like $\Lambda Z_{(1)}$ except that the (even, odd) gradation does not in general correspond to the algebraic one of (generators, relations).

A common feature of all these spaces is that all Massey products vanish - we'll have more to say about this later.

3. Perturbations

To study other spaces, our deus ex machina or rather the machina itself is a model for $A^*(X)$ obtained from the bigraded model $(\Lambda Z,d) \xrightarrow{\rho} (H(X),0)$ by perturbing d to a different D. By a perturbation, I mean D will be of the form

$$D = d + d_2 + d_3 + \dots$$

where d_i lowers the bottom degree by i[5].

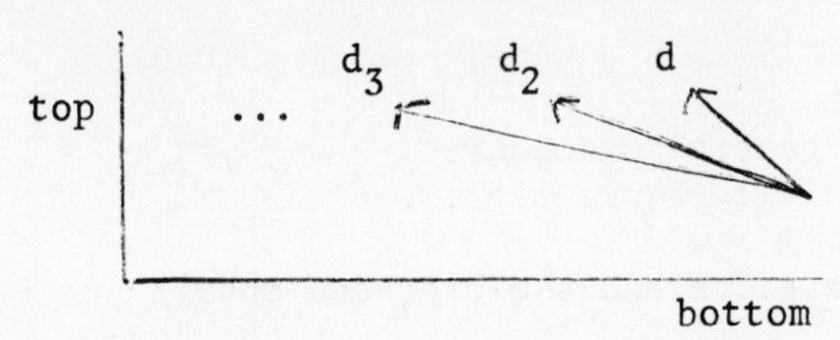

In other words, $(\Lambda Z,D)$ will not be bigraded but rather a filtered (by bottom degree) c.g.d.a.

Before proceeding with the general theory, we study a non-trivial example.

Let S be a space of the form $S^2 \vee S^2 \cup e^5$. The possibilities are determined by $\pi_4(S^2 \vee S^2) \otimes Q \simeq Q \oplus Q$ but let us stay within our context.

$$H(S) \simeq Q[x,y] \otimes E(z)/< x^2,xy,y^2,xz,yz >$$

so the bigraded model looks as follows:

Z_0 is generated by x,y,z, $|x|=|y|=2$, $|z| = 5$,

Z_1 is generated by u,v,w,a,b

with d given by $\begin{cases} du = x^2 & da = xz \\ dv = xy & db = yz \\ dw = y^2 & \end{cases}$

Z_2 is generated by r,s and more

with $dr = xv - yu$

$ds = xw - yv$,

so a perturbation could look like

$$Dr = dr + \lambda z$$

$$Ds = ds + \mu z$$

Notice 1) λ or $\mu \neq 0 \Rightarrow (\Lambda Z,D)$ not minimal

2) $\lambda \neq 0$ say z is a multiple of the Massey product $< x,x,y >$ and similarly for $\mu \neq 0$ and $< x,y,y >$.

As noticed by Body and Douglas [1], there is basically one rational homotopy type other than $S^2 \vee S^2 \cup S^5$, namely $(\lambda,\mu) \neq (0,0)$.

Now let us look at the general procedure.

THEOREM 3. Given (A,d_A) a c.g.d.a and the bigraded model

$$(\Lambda Z,d) \xrightarrow{\rho} (H(A),0)$$

for $H(A)$, there exists a c.g.d.a.

$$(\Lambda Z,D) \xrightarrow{\pi} (A,d_A)$$

which is

1) a cohomology isomorphism which is ρ *in* bottom degree 0, and

2) $D-d$ decreases bottom degree by at least 2.

If π': $(\Lambda Z,D') \rightarrow (A,d_A)$ is another such, then there exists an isomorphism

$$\phi:(\Lambda Z,D) \rightarrow (\Lambda Z,D')$$

such that $\phi-1$ decreases the bottom degree and $\pi'\phi$ is homotopic to π.

For this, of course, we need to define homotopy of c.g.d.a. maps, which is best done by building a model for $(\Lambda Z,D)^I$ which in turn is best done in private [2, 10].

We will concentrate on the construction of π.

Fix a linear splitting $H(A) \xrightarrow{\eta} \Lambda Z_0$ i.e. $\rho\eta = 1$.

Define $D = d = 0$ on ΛZ_0 and let $\pi:Z_0 \rightarrow A$ choose cocycle representatives for $QH(A)$.

Define $D = d$ on Z_1 and for $z \in Z_1$ define π: $Z_1 \rightarrow A$ by choosing $\pi(z_1)$ so that $d_A\pi z_1 = \pi dz_1$ which is possible since $H(\Lambda Z,d) \cong H(A,d_A)$.

Now if $z \in Z_2$, then $dz \in \Lambda Z_0 \otimes \Lambda Z_1$ and $Ddz = ddz = 0$ so $d_A\pi dz = 0$. Define $Dz = dz-\eta[\pi dz]$ so that πDz is a cocycle and hence $\pi(z)$ can be chosen to satisfy $d_A\pi z = \pi Dz$.

Now proceed by induction: Assume D on $\Lambda Z_{(n)}$ and $\pi:(\Lambda Z_{(n)},D) \rightarrow (A,d_A)$.

For $z \in Z_{n+1}$, Ddz is a D-cocycle of bottom degree at most n-2.

LEMMA. For any D-cocycle u of bottom degree at most n-2, there exists v of bottom degree at most n-1 and $\alpha \in H(A)$ such that $u = Dv+\eta(\alpha)$.

Grant this for the moment and apply it to Ddz. Then applying π, we see

$$d_A \pi dz = d_A \pi v + \pi\eta\alpha$$

so that $\pi\eta\alpha$ is homologous to 0, i.e. $\alpha = 0$ and $D(dz-v) = 0$.

Define $Dz = dz-v-\eta[\pi dz-v)]$ then $[\pi Dz] = 0$ so $\pi Dz = d_A b$ for some b and we define $\pi z = b$. The uniqueness result is done similarly - constructing ϕ and the homotopy simultaneously.

Proof of Lemma: Write $u = \sum_{0}^{n-2} u_j$, u_j of bottom degree j. Assume $n-2 \geq 1$. Since $Du = 0$ and du_{n-2} is the only term of deg n-1, it must be 0. But then $u_{n-2} = dv_{n-1}$ since $H_+(\Lambda Z,d) = 0$. Now consider $u-Dv_{n-1}$ which is of lower degree at most n-3 and $D(u-Dv_{n-1}) = 0$. By induction, then, $u-Dv_{n-1} = Dv'+\eta(\alpha)$; take $v = v_{n-1}+v'$.

Now to realize $(\Lambda Z,D)$, just as for $(\Lambda Z,d)$ build successive fibrations $E_{n+1} \to E_n$. The only difference is the inefficiency of the description of the homotopy groups. Where π_* of the formal space is just dual to $\oplus Z_n$, the space realizing $(\Lambda Z,D)$ has homotopy groups given by $H(\oplus Z_n, D|)$ where $D|$ is most easily expressed by regarding $\oplus Z_n$ as the indecomposables of ΛZ.

Thus we have a correspondence between perturbations of $(\Lambda Z,d)$ and spaces X with $H^*(X) \approx H(\Lambda Z,d)$. To study the set of homotopy types realizing a fixed c.g.a. H, we can study the set of perturbations up to the appropriate equivalence relation. The set of perturbations is an algebraic variety and we will see the equivalence is also algebraic.

4. OBSTRUCTION THEORY

Before doing this, we need to look at the following problem.

Given c.g.d.a's (A,d_A), (B,d_B) and $f\colon\ H(A) \simeq H(B)$, does there exist $\phi\colon\ A \to B$ such that $\phi^* = f$?

Notice that if $\rho\colon\ (\Lambda Z,d) \to (H(A),0)$ is the bigraded model, then $\phi\rho$ is the bigraded model for $H(B)$, and so there is a perturbation D_B making $(\Lambda Z,D_B)$ the filtered model for B. Thus the problem of realizing ϕ by an equivalence $f\colon\ A \to B$ can be "reduced" to trying to construct an automorphism of ΛZ which is a chain map from D_A to D_B and of the form 1 plus terms which lower degree.

It may be of either motivational or historical interest to know that Halperin and I started with the special case:

When is a space formal? i.e. when is $(\Lambda Z,D)$ equivalent to $(\Lambda Z,d)$?

For this there was a folk medecine prescription - iff all Massey products vanish - where "vanish" needed to be interpreted very carefully. For $S^2 \vee S^2 \cup e^5$, this said that only $S^2 \vee S^2 \vee S^5$ was formal. The more general problem needed to be solved to see all non-formal $S^2 \vee S^2 \cup e^5$ were rationally homotopy equivalent to each other. Hence the Massey product is a complete invariant.

For as simple an example as $(S^2 \vee S^2) \times S^3$ which is formal, the vanishing of Massey products criterion can be very misleading since the indeterminacy is all of H^5. Fortunately the machineary we have set up avoids this language and its problems.

<u>Obstruction theory</u>

Recall we have the following machinery at our disposal:

1) bigraded (minimal) models:

$$\begin{array}{ccc} (\Lambda Z,d) & \longrightarrow & H(A) \\ & \searrow & \downarrow \phi \\ \rho' = \phi\rho & & H(B) \end{array}$$

(By the uniqueness theorem $\phi\rho$ exhibits $(\Lambda Z,d)$ as the bigraded model for $H(B)$.)

2) filtered models:

$$\begin{array}{c} (\Lambda Z,D_A) \xrightarrow{\pi} A \\ \psi \downarrow \\ (\Lambda Z,D_B) \xrightarrow{\pi'} B \end{array}$$

and we seek ψ inducing ϕ, with ψ-1 decreasing filtration.

Our first move is forced: $\Lambda Z_0 \longrightarrow \Lambda Z_0$ must be the identity and we can extend the identity to ΛZ_1 since $D_A = D_B = d$ there.

So assume ψ: $(\Lambda Z_{(n)}, D_A) \to (\Lambda Z_{(n)}, D_B)$ such that ψ-1 decreases filtration. Call ϕ an n-realizer.

Define the obstruction $0(\phi)$: $Z_{n+2} \to H(B)$ by

$$0(\phi)(z) = [\pi_B \ \phi \ D_A Z]$$

Thus $0(\phi) \in \text{Hom}^1 (Z_{n+1}, H(B))$. If A is of finite type, this can be identified with a subset of

$$\bigoplus_p H^{p+1}(B; \pi_p(H(A))$$

just where obstructions in topology usually lie.

On encountering an obstruction, the next step in topology is to back off a step and consider an alternate ϕ'. We are able to consider

$0_{n+1}(f) = \{0(\phi) \mid \phi$ any n-realizer of f $\}$ and still identity it as a reasonable coset.

Consider the space of all graded ΛZ_0 - derivations of $(\Lambda Z,D_B)$:

$$\text{Der } (\Lambda Z,D_B) = \bigoplus \text{Der}\,{}^p_n\,(\Lambda Z,D_B)$$

where the p and n indicate the derivation raises degree by p for the top degree and lowers by n for the bottom degree. We are particularly interested in

$$M_n \subset H^0(\text{Der } \Lambda \ Z_{(n)}, D_B)$$

consisting of the homology classes of all derications of top degree 0 which decrease filtration.

Define γ: $M_n \to \mathrm{Hom}^1(Z_{n+1}, H(B))$ by

$$\gamma(\theta)(Z) = [\pi_B \theta D_B Z]$$

Proposition.

$$0_{n+1}\ (f) = 0(\phi) + \gamma(M_n)$$

The idea of the proof is easy:

1. Any two n-realizers ϕ, ϕ' differ by an automorphism ψ of $(\Lambda Z_{(n)}, D_B)$, i.e. $\phi' = \psi\phi$, such that ψ-1 decreases filtration

2. log ψ represents a class in M_n and $0(\phi') = 0(\phi) + \gamma(\log\psi)$.

Conversely given θ representing a class in M_n, $(\exp \theta)\phi$ is another n-realizer.

Notice that

$$\mathrm{Der} \xrightarrow{\exp} \mathrm{Aut}$$

is well defined if the derivation decreases filtration because the series for exp θ is actually finite on any Z_n and similarly for log ψ if ψ-1 decreases filtration.

In the special case, $D_B = d$, i.e. B is formal, things simplify. Because $(\Lambda Z,d)$ is bi-graded and so is Der and H(Der). Moreover $H(\mathrm{Der}\ \Lambda Z,d) \approx H(\mathrm{Der}(\Lambda Z,d;\ H(B))$ being a homotopy invariant so we can identity $\gamma(M_n)$ with

$$\gamma\lambda\ \mathrm{Hom}^0(Z_n, H(B))$$

where λ extends any ψ as a ΛZ_0-derivation.

Let's see this machinery in operation. Consider $B = A^*((S^2 \vee S^2) \times S^3)$ as promised last time.

$\Lambda Z_0 = Q[x,y] \otimes E(z)$

$\Lambda Z_1 = E(u,v,w)$ with $du = x^2$

$dv = xy$

$dw = y2$

$\Lambda Z_2 = Q[r,s]$ with $dr = xv-yu$
$ds = xw-yv$,

so all perturbations are of the form

$$Dv = dr + \mu_1 xz + \mu_2 yz$$

$$Ds = ds + \sigma_1 xz + \sigma_2 yz$$

It is not hard to check that each of these extends to a full perturbation on all of ΛZ such that $D^2 = 0$.

Notice that dr represents $\langle x,x,y\rangle$ and any D shows $\langle x,x,y\rangle \equiv 0$ modulo $x H^3 + H^3 y$, but we will show $(\Lambda Z, D)$ is not formal unless $\mu_1 = \sigma_2$.

If $\mu_1 - \sigma_2 \neq 0$, it is easy to construct an isomorphism with the special case $\begin{pmatrix}1 & 0\\ 0 & 0\end{pmatrix}$ i.e.

$$Dr = dr + XZ$$

$$Ds = ds$$

so let us compute our obstruction there:

$$\text{Take } \phi = \text{id}: \ \Lambda Z_{(1)} \to \Lambda Z_{(1)}$$

$$O(\phi)(z_1) = [dr + xz] = XZ$$

$$O(\phi)(z_2) = 0$$

Does this homomorphism belong to the indeterminacy, i.e. to

$$\gamma\lambda \ \text{Hom}^o(Z_1, H(B))?$$

z_1 is generated by $u,v,w \in z_1^3$ so write

$$\beta u = \beta_1 z, \ \beta v = \beta_2 z, \text{etc. } \beta_i \in Q.$$

Then we compute

$$\gamma\lambda(\beta)\ r = -\beta_2\ xz + \beta_1\ yz$$

$$s = \beta_3\ xz - \beta_2\ yz$$

which is never of the form $\begin{pmatrix}1 & 0\\ 0 & 0\end{pmatrix}$.

Again if $\mu_1 = \sigma_2$ then the obstruction $o(\phi)$, in terms of the obvious basis, is given by $\begin{pmatrix}\mu_1 & \mu_2\\ \sigma_1 & \mu_1\end{pmatrix}$ which is in the image of $\gamma\lambda$; in fact $\gamma\lambda(\beta)$ for $\beta_1 = \mu_2, \beta_2 = \mu_1, \beta_3 = \sigma_1$. The higher obstructions all vanish because from then on Z_n is at least 4-connected and $H^p(B) = 0$ for $p \geq 5+1$.

Notice two things: the obstructions to equivalence are algebraic so that existence of equivalences correspond to solutions of algebraic equations and for finite complexes (rationally) we have to solve the equations only for a finite number of dimensions (since the connectivity of Z_n increases with n).

5. THE SPACE OF HOMOTOPY TYPES

We have seen that any s.c.c.g.a. H can be realized as $H^*(X;Q)$ for some X and in fact in many cases by more than one X. So far in all our examples, the set of homotopy types of such X has been a very finite set, so perhaps it is worth noting that for $S^2 \vee S^2 \cup e^{q+1}$, there are infinitely many homotopy types as soon as $\pi_q(S^2 \vee S^2) \otimes Q$ is of dimension greater than five since the set of homotopy equivalences between two such spaces is at most $GL(2,Q) \times GL(1,Q)$.

To attempt to classify the homotopy types in general, consider

$$\mathcal{P}_H = \{\text{perturbations of } (\Lambda Z,d) \xrightarrow[\rho]{} (H,0)\}$$

and the quotient map

$$\mathcal{P}_H \to \{\text{homotopy types } X \text{ with } H(X) \approx H\}.$$

First consider the possible perturbations $D = d+d_2+d_3+ \ldots$ where $d_i\colon\ Z_n \to (\Lambda Z)_{n-i}$ lowers bottom degree by i. We can try to construct D inductively either on n or on i. The latter derives from the deformation theory of algebras [12] but survives the obstruction theory there to yield a good global description.

First consider D^2 expanded as $d^2 + (dd_2+d_2d) + (dd_3+d_3d+d_2d_2) + \ldots$ grouping terms of fixed bottom degree. Thus $D^2 = 0$ implies d_2 is a cocycle in the complex Der $(\Lambda Z,d)$. Here for any bigraded c.d.a.A, we let

$$\text{Der } A = \oplus \text{ Der}^p_n A$$

$$\text{Der}^p_n A = \{\text{derivations of A which raise top degree}$$

by p and which lower bottom degree by n}. This inherits a differential d i.e. $d(\theta) = d\,\theta - (-1)^p\theta\, d$. This is the cohomology referred to. The class of d_2 is in H^1_2 (Der $(\Lambda Z,d)$).

Now look at the next term. This says $D^2 = 0$ implies $d_2 d_2 = 0 \varepsilon H_4^2(\mathrm{Der})$. If $d_2 d_2 = 0$, the Massey-Lie product is defined and it vanishes iff there is a choice of d_3 for which a d_4 exists. [The only reference I know to such Massey-Lie products is Douady, Séminarire H. Cartan 1960/61, exposé 4 where it occurs in the context of higher order obstructions to deformation of complex structure.]

Now the above description smacks of a spectral sequence. It is helpful to study the underlying gadget.

Schlessinger pushed me into this by claiming there is a versal deformation, one from which all others can be induced. Theorem. There is a c.g.d.a. R and an R-derivation on $(\Lambda Z, d) \otimes R$ which is a perturbation of $d \otimes 1$ such that perturbations of d are in 1-1 correspondence with projections $\Lambda Z \otimes R \to \Lambda Z$ induced by c.g.d.a. maps $R \to Q$.

The ring R is in fact the standard complex for the cohomology of $Ⓗ = \mathrm{Der}\,(\Lambda Z, d)$ as a d.g.Lie algebra. Recall that $\mathrm{Der}\,\Lambda Z = \bigoplus \mathrm{Der}\,{}^p_n \Lambda Z$ where both n and p can range over all integers, not just non-negative, although we are primarily interested in $p = 1$ and $n \geq 2$.

The usual bracket $[\theta,\phi] = \theta\phi - (-1)^{pq}\,\phi\theta$ ($\theta \varepsilon \mathrm{Der}^p$, $\phi \varepsilon \mathrm{Der}^q$) and differential $d(\theta) = d\theta - (-1)^p\,\theta d$ make $\mathrm{Der}\,\Lambda Z$ a differential graded Lie algebra, call it Ⓗ. The "standard construction" C Ⓗ is the free commutative coalgebra on s^{-1} Ⓗ with a mixed differential $\mathcal{D}$. Indeed C Ⓗ is (additively) isomorphic to $\Lambda\, s^{-1}$ Ⓗ, in fact being degreewise dual to $\Lambda\, s^{-1}$ Ⓗ* where Ⓗ* = Hom (Ⓗ, Q). The differential $\mathcal{D}$ is defined by extending $d + \frac{1}{2}[\ ,\ ]$ in the obvious way. For example

$$\mathcal{D}(\theta) = d(\theta)$$

$$\mathcal{D}(\theta \wedge \phi) = d(\theta) \wedge \phi \pm \theta \wedge d\phi + 1/2\, [\theta,\phi]$$

Define the homology

$$H^{DGL}(\textcircled{H}) = H(\mathcal{C}\textcircled{H}, \mathcal{D})$$

and the cohomology

$$H_{DGL}(\textcircled{H}) = H(\mathrm{Hom}(\mathcal{C}\textcircled{H}, Q), \mathcal{D}^*)$$

Now a perturbation D of d almost induces a coalgebra map $\chi_D: Q \to \mathcal{C}\textcircled{H}$ by $\chi_D(1) = 1 + s^{-1}p + s^{-1}p \wedge s^{-1}p + s^{-1}p \wedge s^{-1}p \wedge s^{-1}p + \ldots$ (For this to be precise we must use the completion of $\mathcal{C}\textcircled{H}$ or equivalently the double dual $(\mathcal{C}\textcircled{H})^{**}$) so that

$$\chi_D^*: (\mathcal{C}\textcircled{H})^* \to Q$$

by extending multiplicatively $\chi_D^*(h) = \Sigma\, h_i(d_i)$ where h_i: $\mathrm{Der}_i\, \Lambda Z \to Q$ and $p = \Sigma\, d_i$. The fact that $D^2 = 0$ is equivalent to χ_D being a chain map.

The use of the Lie structure on $\textcircled{H} = \mathrm{Der}\, \Lambda Z$ is quite natural, especially in light of our earlier use of $\mathrm{Der}\, \Lambda Z$ in the obstruction theory. It leads however to some rather forbidding formulas if carried out explicitly. An alternative is to pass to the associated universal enveloping (associative) algebra $U\textcircled{H}$ and take its homology as a d.g.a, indeed:

$$H^{DGL}(\textcircled{H}) \overset{=}{} H^{DGA}(U\textcircled{H})$$

As Douady says: "On peut chercher à calculer ... sans sortir du $\textcircled{H}$; les calculs sont bein plus compliqués". That is the formulas are much neater in terms of $U\textcircled{H}$ and more familiar as far as Massey products are concerned. On the other hand $\Lambda s^{-1}\textcircled{H}^*$ being a c.g.da. has a topological significance compatible with our point of view as we shall soon see.

We have "parameterized" the space of perturbations $\mathcal{P}_H$, but need to investigate the relation of homotopy equivalence. Since the conference, Schlessinger and I have been able to show that $D = d+p$ gives the same homotopy type as $D' = d+q$ keeping H fixed if and only if there is a path from $\mathcal{X}_D$ to $\mathcal{X}_D'$, i.e. a map of d.g. algebras

$$R = C(\text{Ⓗ})^* \longrightarrow I$$

where I is the commutative d.g. algebra $Q[t] \otimes E(dt)$. Thus

Theorem. The space of rational homotopy types with fixed isomorphism $i: H(X) \approx H$ is in 1-1 correspondence with the set of path components of $C(\text{Ⓗ})^*$.

The advantage of this result is that it implies that the space of rational homotopy types is a <u>homotopy</u> functor of Der ΛZ. In particular, in nice cases, we might hope to calculate in terms of H(Der ΛZ). If we filter $C(\text{Ⓗ})$ by the Λ-degree, we obtain the spectral sequence referred to earlier:

$$H^{Lie}\, H(\text{Ⓗ}) \Rightarrow E^o\, H^{DGL}(\text{Ⓗ})$$

which is the analog of

$$H^{alg}\, H(G) \Rightarrow E_o H(BG).$$

One would expect the spectral sequence to collapse if $(\Lambda Z, d)$ were nice enough, e.g. Z itself dual to a free Whitehead algebra or a trivial Whitehead algebra. Topologically these would be a wedge of spheres and a product of $K(\pi,n)$'s.

Recall that in our model $(\Lambda Z, d)$, the space Z can be indentified with Hom $(\pi_*(X_H), Q)$. Let π be defined so that $s\pi$ = Hom (Z,Q) and π inherits the structure of a graded Lie algebra corresponding to the Whitehead product. Suppose also d is dual to the Whitehead product.

Define $\tilde{\pi} = s\pi \oplus$ Der (π) where Der means derivations of graded Lie algebras this time and is tself a graded Lie algebra. Regard $s\pi$ as an abelian Lie algebra and let Der act on $s\pi$ via the isomorphism

$s: \pi \to s\pi$. Define a differential δ on $\tilde{\pi}$ by

$$\delta | \mathrm{Der}\,(\pi) = 0$$

$$\delta | s\pi \approx \mathrm{ad}: \pi \to \mathrm{Der}\,(\pi).$$
$$\theta \mapsto [\theta\ ,\ ]$$

Theorem. (Schlessinger): There is a canonical map of d.g. Lie algebras $\zeta: \tilde{\pi} \to \textcircled{H}$. When π is free as a Lie algebra, ζ induces an H_{DGLie} isomorphism.

The map ζ is defined almost tautologically: If $\phi \in \mathrm{Der}\,(\pi)$, then the suspension of the dual ϕ^* maps Z to Z and can be extended as a derivation, $\zeta(\phi) \in H$. Similarly if $\omega \in s\pi$, the dual $\omega^*: Z \to Q$ can be extended to a derivation $\zeta(\omega) \in \textcircled{H}$ That ζ commutes with [,] is a trivial computation and ζ is a chain map from δ to d precisely because d is dual to [,].

If $\pi = L(\sigma)$, the free graded Lie algebra on σ, then by a series of reductions we can identify ζ with the inverse to a map

$$\mathrm{Der}\,(\Lambda Z,d) \to s\pi \oplus (\pi \otimes \sigma^*).$$

Thus $H_{DGLie}(\tilde{\pi})$ is reduced to $H_{Lie}(\mathrm{Der}\,(\pi)/\mathrm{ad}\pi)$ and the path components of $C\,\textcircled{H}$ and $C\ \mathrm{Der}\,(\pi)/\mathrm{ad}\pi$ agree.

In particular if X is a k-fold wedge of S^2's, then σ is a Q-vector space of rank k, all in top degree-1. We have $\mathrm{Der}\,(\pi) \approx \mathrm{Hom}\,(\sigma,\pi)$ is zero if top deg > 0 or when $p \geq 2$ and $n \geq -2$. Thus the portion of $C\,\textcircled{H}$ corresponding to perturbations is zero, and we recover the known result that $H(\vee S^2)$ is intrinsically formal, i.e. is realized by only one rational homotopy type.

However $H^{-\nu}_{\nu}\,\textcircled{H}$ is not zero nor is the corresponding $H^{\nu+1}_{DGLie}$. Schlessinger observed that this gives rise to non-formal homotopy types associated to H in the following way. Theorem. (Schlessinger) Given the

the bigraded model $(\Lambda Z,d) \underset{\rho}{\to} (H,0)$ for the k-fold wedge of S^2's, there is a non-trivial map

$$H^{-2\nu}_{2\nu}(\text{Ⓗ}) \to \{\text{Q-homotopy types realizing } H \otimes H(S^{\nu+1})\}$$

given by $\theta \mapsto d \otimes 1 + \theta \otimes a$ where a is a generator of $H(S^{2\nu+1})$. (The case of $S^{2\nu}$ is only slightly more complicated.)

Several observations are in order:

1) Our example $H((S^2 \vee S^2) \times S^3)$ is of this form.

2) All such differentials correspond to fibrations $W \to E \to S^{2\nu+1}$.

3) Sullivan has proposed as a model for the classifying space of fibrations with fibre of the homotopy type of $(\Lambda Z,d)$ precisely the standard construction $C(\text{Ⓗ})^*$ except that he uses Ⓗ^p_n for $p < 0$ only and in degree $p = 0$ only a carefully chosen subspace of Ⓗ^0_n [10].

Finally, having "calculated" the space of homotopy types X with fixed $i:H(X) \approx H$, we still must factor out by the action of Aut H to obtain the space of homotopy types. Even when we can identify the space of homotopy types completely as in $H((S^2 \vee S^2) \times S^3)$, there are subtleties as reflected in the fact that the space of homotopy types should be regarded as a two point space with the non-trivial non-Hausdorff topology:

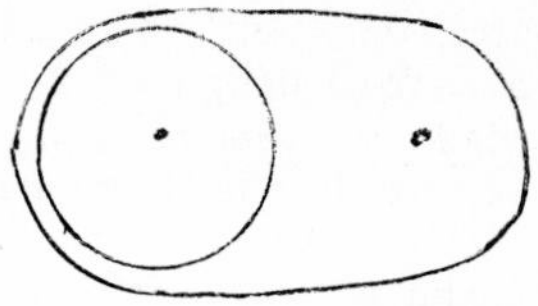

BIBLIOGRAPHY

[1] R.A. Body and R.R. Douglas, Homotopy types within a rational homotopy type, Topology 13 (1974), 209-214.

[2] A.K. Bousfield and V.K.A.M. Gugenheim, "On PL DeRham theory and rational homotopy type", Memoirs of the Amer. Math. Soc. 197, 1976.

[3] P. Deligne, P.Griffiths, J. Morgan and D. Sullivan, The real homotopy theory of Kähler manifolds, Inventiones Math 29(1975) p. 245-254.

[4] J.L. Dupont, Simplicial deRham cohomology and characteristic classes of flat bundles, Topology 15 (1976) 233-245.

[5] V.K.A.M. Gugenheim and J.P. May, On the theory and applications of differential torsion products, Mem. of the Amer. Math Soc. 142, 1974.

[6] S. Halperin and J. Stasheff, Obstructions to homotopy equivalences, Advances in Math (to appear).

[7] P.J. Hilton, On the homotopy groups of a union of spheres, J. London Math. Soc. 30(1955) 154-171.

[8] T. Jozefiak, Tate resolutions for commutative graded algebras over a local ring, Fund. Math. 74 (1972) 209-231.

[9] D. Quillen, Rational homotopy theory, Ann. of Math 90 (1969), 205-295.

[10]. D. Sullivan, Infinitesinal computations in topology, preprint, 1975.

[11] C. Watkiss, Thesis, University of Tornoto, 1976.

[12] M. Gerstenhaber, On the deformation of rings and algebras, I-V, Annals of Math 79(1964) 59-103; 84(1966) 1-19: 88(1968) 1-34; 99 (1974) 257-276; preprint with C. Wilkerson.

Geometric Cobordism and the Fixed Point Transfer

by Albrecht Dold

Introduction. While bordism groups $\Omega_j X$ have immediate geometric appeal the same cannot be said of cobordism groups $\Omega^j X$: The elements of $\Omega_j X$ are represented by maps $M \to X$ of closed j-dimensional manifolds M into X , whereas the elements of $\Omega^j X$ are represented by maps of suspensions of X into Thom-spaces of universal bundles. The situation is similar (although not quite as striking) for ordinary singular homology $H_j X$ versus cohomology $H^j X$. H. Whitney [W], in 1947, proposed a geometric interpretation for $H^j X$ which, however, was not generally adopted - presumably because it was not convenient to use. It was D. Quillen [Q] then, in 1971, who gave a simple geometric interpretation for complex cobordism $U^j X$ of manifolds; he used it to prove deep results in cobordism. I have lectured on Quillen's approach at various occasions (Mexico 1971, [D_2], this conference, a.o.), simplifying it still further, and extending it to more general spaces X and other smooth cobordism theories. Also, it turned out that his approach provides a good understanding of the fixed point index and - transfer of fibre-preserving maps. Therefore, and encouraged by the audiences, I've undertaken to write up these lectures for the present proceedings. An experienced homotopy theorist may not gain new insight from reading these notes, but it is hoped that they will be helpful for younger topologists.

Another geometric approach to cohomology can be found in the recent book [B-R-S]. It applies not only to cobordism but to arbitrary general cohomology $h^j X$ of polyhedra X . As a price for generality, however, much of the appealing simplicity of Quillen's approach is lost here.

The basic section of the present notes is § 2 , where geometric cobordism groups $\Omega_\sigma^j X$ are defined and their elementary properties derived. Section 1 is preliminary; it discusses the various structures σ (orientation, stable complex structure, - parallelisation, etc.) which one imposes on smooth manifolds in order to define (co-)bordism of σ-structured manifolds. The reader is advised to start with § 2 and to think of oriented or stably complex manifolds etc. when the text speaks of σ-manifolds.

In section 3 duality between bordism $\Omega_j X$ and cobordism $\Omega^{n-j} X$ of n-dimensional σ-manifolds X is discussed. This is rather instructive for understanding geometric cobordism or cohomology. Roughly speaking and in terms of singular theory, homology of X is given by finite chains, cohomology by (infinite but) locally finite chains. In (co-)bordism finite chains become maps $W \to X$ of compact manifolds W , whereas locally finite chains become proper maps $W \to X$ of arbitrary manifolds W . (A related bordism functor on proper maps was discussed by Th. Bröcker in unpublished notes entitled "Bordismentheorie auf lokalkompakten Räumen", 1970/71).

Section 4 treats products (cross-, cup-, cap-) in geometric (co-)bordism, and section 5 presents the fixed point index and -transfer in terms of geometric cobordism.

1. Stable homotopy structures on vector bundles

1.1 Definition. We consider the category Vect whose objects are vector bundles $\pi : E \to B$ over smooth manifolds B, (with countable basis, i.e. contained in some R^p), and whose morphisms are bundle maps $f : E \to E'$ (isomorphic on fibres, $f : p^{-1}(b) \cong p'^{-1}(\bar{f}(b))$, where $\bar{f} : B \to B'$ is induced by f).
A contravariant functor

$$(1.2) \qquad \sigma : \text{Vect} \longrightarrow \text{Sets}$$

together with a natural transformation

$$(1.3) \qquad s : \sigma(E) \longrightarrow \sigma(R\times E) \qquad (\text{where } R\times E \to B)$$

is called a structure functor if it has the following properties.

(i) Homotopy invariance (HTP). The projection $q : R\times E \to E$ (viewed as bundle map over $R\times B \to B$) induces bijections $\sigma E \approx \sigma(R\times E)$, for all E.

(ii) Mayer-Vietoris property (MV). If E is a vector bundle over $B = B_1 \cup B_2$, where B_1, B_2 are open subsets of B, and if $u_j \in \sigma E_j$ are such that $u_1 \mid E_1 \cap E_2 = u_2 \mid E_1 \cap E_2$ then there exists $u \in \sigma E$ such that $u_j = u|E_j$. - Here $E_j = \pi^{-1}(B_j)$, and "restriction" $u|E_j$ etc. stands for $(\sigma(\text{inclusion}))(u)$.

(iii) Additivity (ADD). If E is a vector bundle over $B = \oplus_{j=1}^{\infty} B_j$ then the canonical map $\sigma E \longrightarrow \prod_{j=1}^{\infty}(\sigma E_j)$ is bijective.

(iv) Stability (STAB). The transformation s is an equivalence,

$$s : \sigma(E) \simeq \sigma(R\times E) .$$

(For an equivalent notion compare 1.4, example (x).)

(1.4) Examples. (0) $\sigma E = \emptyset$ for all E. This trivial example is usually excluded. If so then the inclusion of the fibre shows that $\sigma(\mathbb{R}^n \to pt) \neq \emptyset$. And ADD shows that the empty bundle (over $B = \emptyset$) has $\sigma(\emptyset) = 1$.

(i) σE consists of one element, i.e. every vector bundle has a unique structure $u \in \sigma E$.

(ii) σE = set of orientations of E, i.e. reductions of the structure group $Gl(\mathbb{R})$ to $Gl^+(\mathbb{R})$.

(iii) σE = set of homotopy classes of stable complex structures on E.

(iv) σE = set of homotopy classes of stable trivializations $(\mathbb{R}^m \times E \cong \mathbb{R}^N \times B)$.

(v) σE = set of h-orientations, where h is a general cohomology theory.

Note: The examples (i)-(iv) amount to (stable) reductions of the structure group; the examples (i),(ii),(iv) are also special cases of (v).

(vi) $\sigma E = \{y \in H^1(B;\mathbb{Z}) \mid y \equiv w_1 E \mod 2\}$.

(vii) If σ_1, σ_2 are structure functors then so is $\sigma E = (\sigma_1 E) \times (\sigma_2 E)$.

(viii) If $\pi : E \to B$ is a vector bundle let $\pi' : E' \to B$ denote the inverse bundle. It is obtained by embedding E into a trivial bundle $\mathbb{R}^n \times B$ and taking the quotient bundle, $E' = \mathbb{R}^n \times B/_E$. This is not unique but if E'' is a second choice (of the same dimension) then there is a canonical homotopy class of isomorphisms $\mathbb{R}^N \times E' \cong \mathbb{R}^N \times E''$, for large N. Given a structure functor σ this allows to define a new (dual) structure functor σ' by $\sigma' E = \sigma E'$. - In many cases (cf. 1.22) σ' is naturally isomorphic to σ.

(ix) $\sigma E = \sigma_k E$ = set of (stable) k-codimensional sub-bundles of E

modulo automorphisms of E .

(x) Let $\gamma : \Sigma \to BO$ a fibration and put $\sigma E = \sigma_\gamma E$ = set of homotopy classes of liftings $\lambda : B \to \Sigma$ of the (stable) classifying map $B \to BO$ of E . This notion was introduced by R. Lashof (cf. [S], Chap.II). Using E.H. Brown's representation theorem [B] one can show that every structure functor σ is equivalent to some σ_γ .

<u>1.5 Comments on the axioms</u>. (0) HTP implies, of course, that $\sigma(f_o) = \sigma(f_1)$, for homotopic bundle maps.

(i) Some natural candidates for σ are equipped with a transformation s which is, however, not an equivalence. In this case one can force stability by replacing σ by the direct limit of

$$\sigma E \xrightarrow{s} \sigma(R\times E) \xrightarrow{s} \sigma(R^2\times E) \xrightarrow{s} \ldots \quad .$$

E.g. this is how one arrives at examples (iii) and (iv).

(ii) The following property is sometimes easier to verify than the M-V-property - but is in fact equivalent to it (assuming homotopy invariance 1.1(i)). Given $u_j \in \sigma E_j$ as in (1.1 ii) we don't insist on finding u such that $u|E_j = u_j$ exactly; we are satsified if we can find $u \in \sigma E$, depending on open subsets $B_j' \subset B_j$ such that $\overline{B_j'} \subset B_j$, with $u|E_j' = u_j|E_j'$. In order to show that this implies the M-V-property one considers the vector bundles $R\times E_j \longrightarrow R\times B_j$ with the induced (by projection) structures, also denoted by $u_j \in \sigma(R\times E_j)$. More precisely, one considers these over the open subsets

$$L_1 = (-\infty,1) \times B_1 \cup R \times(B_1 \cap B_2) \subset R \times B_1$$
$$L_2 = (2,+\infty) \times B_2 \cup R \times(B_1 \cap B_2) \subset R \times B_2 \quad .$$

Over the intersection $L_1 \cap L_2 = R \times(B_1 \cap B_2)$ the two structures agree hence there is $v \in \sigma(R\times E|L_1 \cup L_2)$ which agrees with u_j over L'_j, where $L'_1=(-\infty,1)\times B_1 \cup (-\infty,{}^5/3)\times(B_1\cap B_2)$, $L'_2=(2,+\infty)\times B_2 \cup ({}^4/3,+\infty)\times(B_1\cap B_2)$. Take a continuous $\tau : B \to [1,2] \subset \mathbb{R}$ such that $\tau|B-B_2 = 1$, $\tau|B-B_1 = 2$. Then graph $(\tau) \subset (L_1 \cup L_2)$, and the restriction of v to graph$(\tau) \approx B$ is an element $w \in \sigma(\mathbb{R}\times E|\text{graph}(\tau)) \approx \sigma(E)$, which agrees over B_j with u_j (because $b_j \mapsto (\tau(b_j),b_j)$ for $b_j \in B_j$, is homotopic in $L_1 \cup L_2$ to $b_j \mapsto (j,b_j)$.

(iii) The reader will notice that for most purposes finite additivity would be enough. It seems, however, that 1.20 and the duality theory in section 3 do require ADD, i.e. countably-infinite additivity.

(iv) If A is a locally compact (equivalently: locally closed) subset of a manifold B then every vector bundle E over A extends to a neighborhood in B, and any two extensions are isomorphic (rel E) in a neighborhood (cf.[A]). This allows to speak of σ-structures on vector bundles over A. We omit the details.

(1.6) Opposite structure. If $E \to B$ is a vectorbundle, then so is $\mathbb{R}\times E \to B$, and the map $\nu\times\text{id} : \mathbb{R}\times E \to \mathbb{R}\times E$, where $\nu(t) = -t$ is a bundle map over id_B. It induces an involution $\sigma(\nu\times\text{id})$ of $\sigma(\mathbb{R}\times E) = \sigma(E)$, called passage to the opposite structure. If $u \in \sigma E$ we also write u^- for its opposite, $u^- = (\sigma(\nu\times\text{id}))(u)$.

The passage to the opposite is compatible with stability: $s(u)^- = s(u^-)$. This amounts to comparing the two bundle maps $\nu \times \text{id} \times \text{id}$, $\text{id} \times \nu \times \text{id} : \mathbb{R} \times \mathbb{R} \times E \to \mathbb{R} \times \mathbb{R} \times E$. But they are homotopic as bundle maps, and therefore induce the same involution on $\sigma(\mathbb{R}\times\mathbb{R}\times E)$.

1.7 Definition. A smooth manifold X is structured by structuring its tangent bundle $TX \to X$: i.e. a σ-structured manifold is a pair

(X,u) where X is a smooth manifold and $u \in \sigma(TX)$.

<u>1.8 Elementary properties</u>. If X is structured then every open subset $Y \subset X$ is structured (induced by inclusion); isotopic open embeddings $i_o, i_1 : Y \longrightarrow X$ induce the same structure on Y . The tangent bundle of $\mathbb{R} \times X$ is $(\mathbb{R}\times\mathbb{R}) \times (TX) \longrightarrow \mathbb{R}\times X$; its structures are in 1-1 correspondence with those of $TX \to X$; structuring the manifold $\mathbb{R} \times X$ (or $\mathbb{R}^n \times X$) is therefore equivalent to structuring X . Structures on $X_1 \oplus X_2$ are pairs of structures. If X_1, X_2 are structured open subsets of X which induce the same structure on $X_1 \cap X_2$ then $X_1 \cup X_2$ admits a structure which extends the given structures on X_1, X_2 .

Every $\mathbb{R}^n$ can be structured (unless $\sigma \equiv \emptyset$) : if we fix one structure on $\mathbb{R}$ (or on $\mathbb{R}^o = pt$) then <u>every $\mathbb{R}^n$ and every open subset X of $\mathbb{R}^n$ is canonically structured</u>.

1.9 <u>Further Comment on the axioms</u>. In the following sections we shall need structured manifolds much more than structured vector-bundles; in 1.8 we've listed their basic properties. It would therefore be natural to use these properties for an axiomatic approach. This would have the further advantage of being more geometric than 1.1. However, the axioms 1.1 on vector-bundles are easier to use; proofs become shorter. Moreover, the two approaches are, in fact, equivalent: We've already indicated how to pass from structured vector-bundles to structured manifolds. For the inverse process one uses the normal bundle $T'X \to X$ of manifolds, i.e. the open tubular neighborhood of a smooth embedding $X \subset \mathbb{R}^p$. Structuring a vector-bundle $E \to X$ then amounts to struturing the manifold $E \times_X (T'X)$, i.e. the total space of the Whitney-sum $E \oplus T'X$.

We shall not carry out the details of this equivalence, as it is not needed here.

<u>1.10 Definition</u>. A continuous map $f : Y \to X$ between smooth manifolds is σ-structured by structuring the bundle $TY \oplus f^*T'X$, where $T'X$ is the normal bundle of X with respect to a smooth imbedding; equivalently, $T'X$ is the inverse of the tangent-bundle, $TX \oplus T'X = R^N \times X$. This is well-defined because $T'X$ is well-defined up to stable equivalence. For instance, if $(f : Y \to X) = (\pi : E \to X)$ is itself a vector-bundle then $TY = TE = \pi^*E \oplus \pi^*TX$, hence $TY \oplus f^*T'X = \pi^*E \oplus \pi^*TX \oplus \pi^*T'X = \pi^*E \oplus \pi^*(TX \oplus T'X) = \pi^*E \oplus \text{trivial bundle} = \pi^*E \times R^n$. It follows that there is a canonical 1-1 correspondence between σ-structures of the continuous map $\pi : E \to B$ and σ-structures of the vector-bundle $\pi : E \to B$; we can therefore ignore the distinction. The <u>set of σ-structures of f will be denoted by $\sigma(f)$</u>.

<u>1.11 Examples and elementary properties.</u> Constant mappings of X (e.g. $X \to pt$) or nullhomotopic mappings (e.g. $X \to R^n$) are structured by structuring the vector-bundle $TX \to X$, i.e. by structuring X. The identity mapping $id : X \to X$ is structured by structuring $TX \oplus T'X$, i,e. by structuring the trivial vector-bundle $R^n \times X \to X$; in particular, $id : X \to X$ has canonical structure (corresponding to $u \in \sigma(R)$). If $\pi : E \to X$ is a vectorbundle with zero-section $\iota : X \to E$ then $\sigma(\iota : X \to E) \approx \sigma(\pi' : E' \to X)$, where E' is the inverse bundle; this follows because $\iota^*(T'E) = E' \oplus T'X$. If $f : Y \to X$ is any map and V is an open neighborhood of $f(Y)$ in X then $\sigma(f : Y \to X) = \sigma(f : Y \to V)$.

If $f : Y \to X$ is a smooth fibre bundle then $TY = (\tau f) \oplus (f^*TX)$, where $\tau f = \ker(Tf : TY \to TX)$ is the "bundle of tangent vectors along the fibre". It follows that

$$TY \oplus (f^*T'X) = (\tau f) \oplus f^*(TX \oplus T'X) = (\tau f) \oplus \text{trivial},$$ hence

$\sigma(f) = \sigma(\tau f)$. <u>$\sigma$-structuring a bundle-projection therefore amounts</u>

to σ-structuring the bundle along the fibre. More generally, the same arguments apply to (smooth) submersions (while immersions, just as $\iota : X \to E$ above, are structured by structuring the inverse of the normal bundle).

(1.12) Induced structure. Two smooth maps $\alpha : Y \to X \leftarrow M : g$ are said to be transverse if the joint derivative $(T_\eta\alpha, T_\mu g) : T_\eta Y \oplus T_\mu M \longrightarrow T_\xi X$ is surjective for every pair $(\eta,\mu) \in Y \times M$ such that $\alpha(\eta) = \xi = g(\mu)$. Equivalently, (α,g) are transverse if $\alpha \times g : Y \times M \longrightarrow X \times X$ is transverse to the diagonal in the usual sense (cf. [H],3.2.1). If (α,g) are transverse then $N = Y \times_X M = (\alpha\times g)^{-1}(\text{diagonal})$ is a smooth submanifold of $Y \times M$, and we have the commutative pullback-diagram of smooth maps

$$\begin{array}{ccc} Y \times_X M = N & \xrightarrow{\alpha'} & M \\ \downarrow{\scriptstyle g'} & & \downarrow{\scriptstyle g} \\ Y & \xrightarrow{\alpha} & X \end{array} \tag{1.13}$$

We want to define $\sigma(g) \longrightarrow \sigma(g')$, as induced by 1.13. We use the sequence

$$TN \overset{(Tg',T\alpha')}{\rightarrowtail} (TY)_N \oplus (TM)_N \overset{(T\alpha,-Tg)}{\twoheadrightarrow} (TX)_N ,$$

where $(TX)_N = (\alpha g')^*TX$ etc. denotes the induced vector bundles over N. The sequence is easily seen to be exact. It splits (as always for vector-bundles) in a unique way up to homotopy, hence $(TX)_N \oplus TN \cong (TY)_N \oplus (TM)_N$. If we add $(T'X)_N \oplus (T'Y)_N$ on both sides and use $TX \oplus T'X = R^p \times X$, $TY \oplus T'Y = R^q \times Y$ we obtain

$$R^p \times (TN \oplus (T'Y)_N) \cong R^q \times ((TM)_N \oplus (T'X)_N) ,$$

hence $\sigma(TN \oplus (T'Y)_N) = \sigma((TM \oplus (T'X)_M)_N)$. The bundle map $(TM \oplus (T'X)_M)_N \longrightarrow TM \oplus (T'X)_M$ over α' therefore assigns to every σ-structure $u \in \sigma(TM \oplus (T'X)_M) = \sigma(g)$ a σ-structure

$u' \in \sigma(TN \oplus (T'Y)_N) = \sigma(g')$, as desired. This u' (up to sign; cf. 1.15) is called the induced (from u , by α) structure.

1.14 Corollary. If α (in 1.13) is a homotopy equivalence then $\sigma(g) \longrightarrow \sigma(g')$ is bijective - because then α' is also a homotopy equivalence, hence $\sigma(E) \approx \sigma(E_N)$ for every vector-bundle over M ; apply this to $E = TM \oplus (T'X)_M$. ∎

1.15 Remark. Signs are needed to make the construction $\check{\alpha} : \sigma(g) \longrightarrow \sigma(g')$ above functorial in α , i.e. $(\alpha_1\alpha_2)^{\vee} = \check{\alpha}_2\, \check{\alpha}_1$. We shall need this only in special cases (2.16, 2.18) , and shall give the necessary details there. The general case is left to the reader (the adequate sign seems to be $(-1)^{pq+p}$) .

1.16 Multiplicative structure functors. Many structure functors (examples 1.4 i-vi) can be equipped with a natural multiplication

$$(1.17) \qquad (\sigma E_1) \times (\sigma E_2) \xrightarrow{\ \bullet\ } \sigma(E_1 \times E_2) ,$$

resp. (if the bundles are over the same base)

$$(1.17') \qquad (\sigma E_1) \times (\sigma E_2) \xrightarrow{\ \circ\ } \sigma(E_1 \oplus E_2) ;$$

the latter (1.17') is obtained from (1.17) by applying σ to the diagonal map. We say that σ , equipped with $\bullet$, is a multiplicative structure functor, if the multiplication is associative, commutative (in the graded sense)*), and has a neutral element $1 \in R^0$ (thus $1 \cdot u = u$ for all E and $u \in \sigma E$) . Moreover, the multiplication in $\sigma(R^0)$ is assumed to have inverses, hence $\sigma(R^0)$ is a (multiplicative) abelian group, which operates naturally on σE .

One can show (compare [D_3],6,6) that all elements of σ

---

*) Exchanging factors $E_1 \times E_2 \to E_2 \times E_1$ takes $a \cdot b$ into $b \cdot a$ or $(b \cdot a)^-$, depending on the parity of $(\dim E_1)(\dim E_2)$.

have inverses (compare 1.20) in particular, multiplication with $a \in \sigma(R)$ is bijective, $\sigma(E) \cong \sigma(R \times E)$. We can and shall, therefore assume that the stability isomorphism s coincides with $s(1)\cdot$; thus

$$(1.18) \qquad s : \sigma E \longrightarrow \sigma(R \times E) \ , \ s(u) = s(1)\cdot u \ .$$

It follows that s commutes with the operation of the group $\sigma(R^o)$: $s(\lambda\cdot u) = s(1)\cdot\lambda\cdot u = \lambda\cdot s(1)\cdot u = \lambda\cdot s(u)$.

Passage to the opposite structure (1.6) is induced by $\nu \times id : R \times E \to R \times E$. This is a product map, and multiplication 1.17 is natural, hence $(\sigma(\nu \times id))(a\cdot b) = \sigma(\nu)(a)\cdot b = a^-\cdot b$ for $a \in \sigma R$, $b \in \sigma E$. Therefore,

$$s(1)\cdot u^- = s(u^-) \overset{1.6}{=\!=} (\sigma(\nu \times id))s(u)$$

$$= (\sigma(\nu \times id))(s(1)\cdot u) = s(1)^-\cdot u = s(1^-)\cdot u$$

$$= s(1)\cdot 1^-\cdot u \ , \text{ hence}$$

$$(1.19) \qquad u^- = 1^-\cdot u \ ,$$

ie.. the opposite structure u^- is obtained by multiplying u with the opposite structure 1^-.

Since $(u^-)^- = u$ we have $1^-\cdot 1^- = 1$, thus $1^- \in \sigma(R^o)$ has order at most 2.

Just as with homotopy classes of maps into H-spaces we have the following

1.20 Proposition. If E_1, E_2 are vectorbundles over B and $u_1 \in \sigma E_1$, $u \in \sigma(E_1 \oplus E_2)$ are arbitrary structures then there is a unique structure $u_2 \in \sigma E_2$ such that $u = u_1\cdot u_2$. In other words, the mapping

(1.21) $\qquad (\sigma E_1) \times (\sigma E_2) \longrightarrow (\sigma E_1) \times \sigma(E_1 \oplus E_2)$,

$$(u_1, u_2) \longmapsto (u_1, u_1 \cdot u_2)$$

<u>is bijective for all E_1, E_2 . - Similarly for the exterior product in $\sigma(E_1 \times E_2)$. -</u>

For our examples in 1.4 this proposition is easy (i-iv and vi) , or well-known (v; cf. [D_3], 6.6). The general proof is lengthy and would detract from the intention of these notes; moreover, the proposition or its corollaries will only be used much later (not in section 2). We shall therefore omit the proof. If one admits the more familiar representability result mentioned under 1.4 (x), i.e. if one knows that σE coincides with the set of homotopy classes of liftings λ of the classifying map

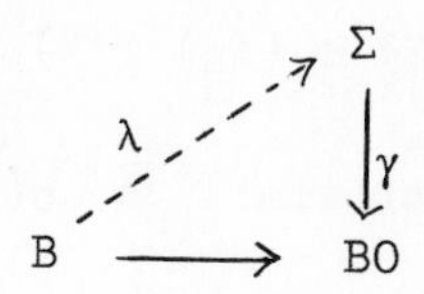

then the proof becomes easy (using the fact that Σ is an H-space and γ an H-map). ∎

If $E_2 = E'$ is the inverse bundle of $E = E_1$ then $E_1 \oplus E_2 = R^n \times B$ and we have the canonical structure $s^n(1)$ on (R^n and hence on) $R^n \times B$. Thus

<u>1.22 Corollary</u>. <u>For every structure $u \in \sigma(E)$ there is a unique structure $u' \in \sigma(E')$ on the inverse bundle E' s.t. $u \cdot u' = s^n(1) \in \sigma(R^n \times B) = \sigma(E \oplus E')$</u> ; this u' is called the inverse structure. <u>The correspondence $u \mapsto u'$ is a natural equivalence between σ and the dual structure functor σ'</u> . ∎

<u>1.23 Corollary</u>. <u>The set $\sigma(id:B \to B)$ of structures on the trivial</u>

0-dimensional vector-bundle over B is an abelian group under multiplication 1.17'. For every vector-bundle $E \to B$ this group $\sigma(id_B)$ operates simply transitively on σE (thus $\sigma E = \emptyset$, or $\sigma E \approx \sigma(id_B)$). - Indeed, the first part follows from 1.20 if we take $E_1 = E_2 = B$, the second part with $E_1 = E$, $E_2 = B$. ∎

1.24 Structuring composite maps. We still assume a multiplicative structure functor σ. Consider continuous maps $Z \xrightarrow{g} Y \xrightarrow{f} X$ between smooth manifolds. Recall (1.10) that f is structured by structuring the bundle $TY \oplus f^*T'X$, in symbols $\sigma(f) = \sigma(TY \oplus f^*T'X)$. Consider the maps

$$\sigma(g) \times \sigma(f) = \sigma(TZ \oplus g^*T'Y) \times \sigma(TY \oplus f^*T'X) \xrightarrow{id \times g^*}$$

$$\sigma(TZ \oplus g^*T'Y) \times \sigma(g^*TY \oplus g^*f^*T'X) \xrightarrow{\cdot}$$

$$\sigma(TZ \oplus g^*(TY' \oplus TY) \oplus (fg)^*T'X) =$$

$$\sigma(TZ \oplus (fg)^*T'X \oplus \text{trivial}) = \sigma(fg) .$$

The composite map, i.e.

$$(1.25) \quad \sigma(g) \times \sigma(f) \longrightarrow \sigma(fg), \ (v,u) \longmapsto u \circ v = v \cdot g^*u$$

defines composition of structures; $u \circ v$ is called the composite structure. It is easy to see that this composition is associative and has neutral elements $u \circ 1_Y = u$, $1_X \circ u = u$ where $1_X \in \sigma(id_X)$ is the canonical element (= neutral element of the group $\sigma(id_X)$).

We can now form the category of structured maps $\mathcal{C}_\sigma$. Its objects are smooth manifolds, its morphisms are structured maps (f,u), where $f : Y \to X$ is as above and $u \in \sigma(f)$. Composition was just defined. The forget-functor $\Phi : \mathcal{C}_\sigma \to \mathcal{C}$ into the category $\mathcal{C}$ of continuous maps between smooth manifolds ($\Phi(f,u) = f$) has some remarkable properties. For instance, 1.20 implies that (f,u) is isomorphic iff f is isomorphic (= diffeomorphic). If $\Phi^{-1}(f) \neq \emptyset$ (i.e. if f admits a structure u) then $\Phi^{-1}(g) \approx \Phi^{-1}(fg)$ (via $v \mapsto u \circ v$) . These and other properties are interesting but not needed here; we omit the details. We remark, however, that the functor Φ might provide an adequate way to axiomatize multiplicative structures, i.e. one would start with some $\Phi : \mathcal{C}' \to \mathcal{C}$ imposing suitable axioms, and define structures by $\sigma(f) = \Phi^{-1}(f)$ etc.

§ 2 Geometric Cobordism

The purpose of this section is to develop cobordism theory $\Omega_\sigma^*(X)$ of σ-structured manifolds in an elementary geometric way, without using Thom spectra or related notions from homotopy theory. The main idea is from Quillen's paper [Q] in which he describes this kind of approach for complex cobordism. We begin by assuming X to be open in some R^p ; later on we admit arbitrary locally compact X which embed into some R^p .

<u>(2.1) The groups $\Omega_\sigma^j(X,A)$ for open subsets $X \subset R^p$.</u> Let $X \subset R^p$ an open subset. A <u>proper σ-manifold</u> over X is a continuous proper map $g : M \to X$, where M is a σ-structured (smooth) manifold. Manifolds $M,N, \ldots$ are usually assumed to have empty boundary, $\partial M = \emptyset$; they will be <u>non-compact</u>, in general. If a manifold is allowed to have non-empty boundary we shall say so , and we shall then use notations $W,W' \ldots$.

Two proper σ-manifolds $g_i : M_i \longrightarrow X$ $i = 0,1$, are said to be cobordant, $g_o \sim g_1$, if a proper σ-manifold $G : W \to X$ exists such that $\partial W \approx M_o \oplus M_1^-$ and $G|\partial W = (g_o,g_1)$. Here $\approx$ means σ-diffeomorphic, i.e. by a diffeomorphism which preserves the σ-structure; the notation M_1^- indicates M_1 with the opposite structure (1.6). Cobordism $\sim$ is an equivalence relation: $W = [0,1] \times M \to M \to X$ shows reflexivity, W^- shows symmetry, and glueing $W \underset{N}{\sqcup} W'$ along a common boundary $N = \partial W \approx \partial W'$ shows transitivity; note that the MV-property (1.1 ii) allows to glue σ-structures (along with the underlying manifolds). Equivalence classes under $\sim$ (cobordism classes) are denoted by $[g]=[g:M \to X]$.

Proper σ-manifolds over X can be added by taking the topological sum (disjoint union); thus, $(g_1,g_2) : M_1 \oplus M_2 \longrightarrow X$. This addition (of σ-diffeomorphism classes) is associative, commutative, with zero-element $\emptyset \to X$ (the empty manifold has a unique σ-structure). Furthermore, it is compatible with $\sim$, i.e. it induces an addition on cobordism classes $[g_1:M_1 \to X] + [g_2:M_2 \to X] = [(g_1,g_2):M_1 \oplus M_2 \to X]$. The relation $\partial([0,1]\times M) \approx M \oplus M^-$ shows $[g:M \to X] + [g:M^- \to M] = [\emptyset]$, hence every $[g]$ has a negative, namely $[g:M^- \to X]$.

For $j \in \mathbb{Z}$, let $\Omega_\sigma^j(X) = \Omega^j X$ denote the set of cobordism classes of proper σ-structured manifolds $M \to X$ of dimension $p - j$, thus $j = \dim X - \dim M$. Under addition + this is an abelian group.

More generally, if $A \subset X$ is a <u>closed</u> subset we consider proper σ-manifolds $g : M \to X$, resp. $W \to X$, whose image lies in $X - A$; thus M resp. W lies over $X - A$ but is proper over X. With these objects $M \to X$, resp. $W \to X$, we form cobordism groups as before which we denote by $\Omega_\sigma^j(X,A)$ where $j = \dim X - \dim M$. Ignoring the fact that the image of g actually lies in $X - A$ defines a homomorphism

$$(2.2) \qquad \varkappa^* : \Omega^j(X,A) \longrightarrow \Omega^j X .$$

If $X' \subset X$ are two open sets in $\mathbb{R}^p$ we define

$$(2.3) \qquad \iota^* : \Omega^j X \to \Omega^j X' , \quad i^*[g:M \to X] = [g':M \to X] ,$$

where $M' = g^{-1}X'$, $g' = g|M'$. Similarly, if $A' \subset X'$, $A \subset X$ are closed subsets with $A' \subset A$ then

$$(2.4) \qquad \iota^* : \Omega^j(X,A) \longrightarrow \Omega^j(X',A')$$

is defined by the same formula.

Clearly ι^* is homomorphic, and it makes Ω_σ^j a contravariant functor on inclusion maps of open subsets of $\mathbb{R}^p$.

(2.5) The groups $\Omega_\sigma^j(X,A)$ for locally compact subsets $X \subset R^p$.

Subsets X of R^p are locally compact iff they are locally closed ([D_1],IV,8.3), i.e. iff they admit an open neighborhood V in R^p such that X is (relatively) closed in V . For such X we define

(2.6) $\Omega_\sigma^j(X) = \varinjlim \left\{ \Omega_\sigma^j(V) \mid V \text{ is an open neighborhood of } X \right\}$.

The direct limit is taken over the direct system of groups $\Omega^j V$ and homomorphisms $\iota^*:\Omega^j V \to \Omega^j V'$. One can, of course, replace the system of all open neighborhoods by any <u>cofinal</u> subsystem. For instance, we can confine ourselves to open neighborhoods V in which X is closed. This shows that we can generalize and define

(2.7) $\Omega_\sigma^j(X,A) = \varinjlim \left\{ \Omega_\sigma^j(V,A) \mid V \text{ open neighborhood of } X \text{ such that } \overline{X} \cap V = X \right\}$

where $X \subset R^p$ is locally compact and A is (relatively) closed in X .

If X is open in R^p then 2.6, 2,7 coincide with the earlier definition in 2.1 because X itself is then an open neighborhood, and it is cofinal, all by itself.

Inclusion maps $(X',A') \subset (X,A)$ of pairs as above induce homomorphisms $\iota^* : \Omega_\sigma^j(X,A) \longrightarrow \Omega_\sigma^j(X',A')$. This is fairly obvious and is left to the reader - all the more so as we shall discuss functoriality again in 2.14.

<u>2.8 Variation on 2.5</u>. The variation consists in describing explicitely, in terms of representatives and equivalences, the elements of $\Omega_\sigma^j(X,A)$ for locally compact $X \subset R^p$ and relatively closed $A \subset X$ - without explicitely mentioning direct limits. This is fairly obvious, using standard facts about $\varinjlim$ (cf [D_1],VIII,5, in particular 5.18).

Elements of $\Omega_\sigma^j(X,A)$ are represented by proper σ-structured manifolds $g : M \to V$, where V is an open neighborhood of X in R^p in which X (or at least A) is closed, and $g(M) \subset (V-A)$. Two representations $g_i : M_i \to V_i$, $i = 0,1$, define the same element of $\Omega_\sigma^j(X,A)$ iff there is a proper σ-structured manifold $G : W \to V$ where $V \subset (V_1 \cap V_2)$ is an open neighborhood of X,
$\partial W \approx (g_o^{-1}V) \oplus (g_1^{-1}V)^-$, $G|\partial W = (g_o,g_1)|\partial W$, and $G(W) \subset (V-A)$.
In particular, if $g : M \to V$ is a representative and $V' \subset V$ is a smaller neighborhood then $M' = g^{-1}V' \to V'$ represents the same element of $\Omega_\sigma^j(X,A)$ - just consider $W = [0,1] \times M'$. Addition in $\Omega_\sigma^j(X,A)$ is represented by taking the topological sum of representatives (over the same V). The homomorphism
$\iota^* : \Omega_\sigma^j(X,A) \longrightarrow \Omega_\sigma^j(X',A')$ is represented by letting $g : M \to V$ represent $a \in \Omega(X,A)$ as well as $\iota^* a$ - at least if A <u>and</u> A' are closed in V; otherwise one uses $g^{-1}V' \to V'$ for $\iota^* a$, with suitable $V' \subset V$.

<u>(2.9) Proposition.</u> <u>The sequence</u>

$$\Omega_\sigma^j(X,A) \xrightarrow{\varkappa^*} \Omega_\sigma^j X \xrightarrow{\iota^*} \Omega_\sigma^j A$$

<u>is exact</u>.

<u>Proof</u>. The elements $a \in \Omega(X,A)$ are represented by proper $g : M \to V$, where V is an open neighborhood of X in R^p, and $g(M) \subset (V-A)$. Since g is proper $g(M)$ is closed in V, hence $V' = V-g(M)$ is an open neighborhood of A in R^p. The element $\iota^* \varkappa^* a$ is represented by $g^{-1}V' \to V'$; but $g^{-1}V' = \emptyset$, hence $\iota^* \varkappa^* a = 0$.

The elements $x \in \Omega X$ are represented by proper $g : M \to V$, where V is an open neighborhood of X in R^p such that $\overline{X} \cap V = X$. If $\iota^* x = 0$ then $g^{-1}V' \to V'$ bounds for some open V'

with $A \subset V' \subset V$, i.e. there is a proper σ-manifold $G : W \to V'$ such that $\partial W \approx g^{-1}V'$ and $G|\partial W = g|g^{-1}V'$. We attach a collar $(\partial W) \times (0,1] \approx (g^{-1}V') \times (0,1]$ to W (along $\partial W \times \{1\}$) and obtain a (non-proper) σ-manifold $W' = W \cup (g^{-1}V') \times 0,1]$ over V. It intersects with $M \times [0,1)$ in $(g^{-1}V') \times (0,1)$, and we can form the union

$$M \times [0,1) \cup W' = M \times [0,1) \cup W .$$

This is over V, and it has a σ-structure (cf. 1.1,(MV)) whose restrictions to $M \times [0,1)$ and W' agree with the given σ-structures.

Now choose a smooth function $\tau : M \times [0,1] \cup W' \longrightarrow [0,1]$ such that

$\tau(z) = 0$ for $z \in (g^{-1}A) \times [0,1) \cup M \times \{0\} \cup G^{-1}A$,

and $\overline{\tau^{-1}[0,1)} \to V$ is proper.

Let $\varepsilon \in (0,1)$ any regular value of τ. Then $\tau^{-1}[o,\varepsilon]$ is a proper σ-manifold over V with boundary $M \times \{0\} \cup M'$, where $M' = \tau^{-1}(\varepsilon)$ lies over $V - A$. Thus $x = [M \to V] = [M' \to V] \in \operatorname{im}(\varkappa^*)$. ∎

<u>(2.10) The coboundary operator $\delta : \Omega_\sigma^j A \to \Omega_\sigma^{j+1}(X,A)$.</u> Suppose first that X is open in R^p (and A relatively closed in X). Elements $a \in \Omega_\sigma^j A$ are then represented by proper σ-manifolds $g : M \to V$ where $V \subset X$ is an open neighborhood of A. Let $\tau : M \longrightarrow [0,1]$ a smooth function such that $\tau|g^{-1}A = 0$, $\overline{\tau^{-1}[0,1)} \to X$ is proper ; and let $\varepsilon \in (0,1)$ a regular value of τ. Then $\tau^{-1}(\varepsilon)$ is a proper σ-manifold over X whose class $[g|\tau^{-1}\varepsilon : \tau^{-1}\varepsilon \to X] \in \Omega^{j+1}(X,A)$ is well defined (independent of ε,τ) ; by definition, this is $\delta(a)$.

Clearly $\tau^{-1}\varepsilon = \partial\tau^{-1}[0,\varepsilon]$, hence $\tau^{-1}\varepsilon \to X$ represents the zero-element of $\Omega^{j+1}X$; thus $\varkappa^*\delta = 0$. Conversely, if $f : N \to X$ represents $b \in \ker(\varkappa^* : \Omega_\sigma^{j+1}(X,A) \longrightarrow \Omega_\sigma^{j+1}X)$ then there exists $g : W \to X$ with $\partial W \approx N$, $g|\partial W = f$. The set $f(N)$ is closed in X and contained in $X - A$, hence $V = X - f(N) = X - g(\partial W)$ is an open neighborhood of A and $M = g^{-1}V \longrightarrow V$ represents an element $a \in \Omega_\sigma^j A$. One easily sees that $b = \delta(a)$. Thus, $\mathrm{im}(\delta) = \ker(\varkappa^*)$.

By similar arguments, $\ker(\delta) = \mathrm{im}(\iota^*)$. Altogether we've now shown

<u>(2.11) Proposition</u>. <u>The sequence</u>

$$\Omega^j(X,A) \xrightarrow{\varkappa^*} \Omega^j X \xrightarrow{\iota^*} \Omega^j A \xrightarrow{\delta} \Omega^{j+1}(X,A) \xrightarrow{\varkappa^*} \Omega^{j+1}X$$

<u>is exact</u>.

We did assume X to be open in some R^p , so far. If X is not open we choose an open neighborhood U of X in R^p such that X (or at least A) is relatively closed in U . Then we define $\delta : \Omega^j A \to \Omega^{j+1}(X,A)$ by composing $\Omega^j A \xrightarrow{\delta} \Omega^{j+1}(U,A) \longrightarrow \Omega^{j+1}(X,A)$. The proposition 2.11 can then be proved (in full generality) by passing to the limit over $\{U\}$ - or by geometric ad hoc arguments as above. ∎

<u>(2.12) Proposition</u> (excision). <u>If X_1, X_2 are locally compact subsets of R^p which are relatively closed in $X_1 \cup X_2$ then</u>

$$\iota^* : \Omega_\sigma^j(X_1 \cup X_2, X_1) \cong \Omega_\sigma^j(X_2, X_2 \cap X_1) .$$

<u>Proof</u>. We argue with representatives over open neighborhoods V of $X_1 \cup X_2$ resp. V_2 of X_2 in R^p , as above.

By choosing them sufficiently small we can assume that $X_1 \cup X_2$

is relatively closed in V, and $X_1 \cap V_2$, X_2 are relatively closed in V_2. Let $\mathcal{O}$ be an open set in R^p such that $\mathcal{O} \cap (X_1 \cup X_2) = X_2 - X_1$.

We now prove surjectivity of ι^*. Elements $b \in \Omega(X_2, X_2 \cap X_1)$ are represented by proper $g : M \to V_2$ with $g(M) \subset (V_2 - X_2 \cap X_1)$. Since $g(M)$ and $X_1 \cap X_2$ are closed in V_2 there is an open neighborhood V_{21} of $X_2 \cap X_1$ in V_2 such that $g(M) \subset (V_2 - V_{21})$. We can restrict $g : M \to V_2$ to smaller neighborhoods of X_2 without changing its class b; in particular, we can assume $V_2 \subset (\mathcal{O} \cup V_{12})$, hence $g(M) \cap X_1 = \emptyset$. Since $g(M)$ and $X_1 \cap V_2$ are closed in V_2 we can find an open neighborhood V_1 of X_1 in R^p such that $g(M) \cap V_1 = \emptyset$. It follows that $g : M \longrightarrow V_1 \cup V_2$ is proper; it represents an element $a \in \Omega(X_1 \cup X_2, X_1)$ such that $\iota^* a = b$.

Injectivity is similar: If $a \in \Omega(X_1 \cup X_2, X_1)$ is represented by $f : N \to V$ (with $f(N) \subset V - X_1$), and $\iota^* a = 0$, then f bounds some $G : W \to V_2$ with $X_2 \subset V_2 \subset V$, $G(W) \subset (V_2 - X_1 \cap X_2)$. Applying to G the same arguments which we used for g before yields a cobordance from f to some $f' : N' \to V'$ whose image avoids X_1 <u>and</u> X_2, hence $a = [f'] = 0$. ∎

2.13 Proposition (continuity). <u>Let $\{(X_\lambda, A_\lambda)\}_{\lambda \in \Lambda}$ a directed (under inverse inclusion) set of pairs in R^p such that X_λ is locally compact and A_λ is relatively closed in X_λ (or all $\lambda \in \Lambda$). Suppose that $X = \bigcap_{\lambda \in \Lambda} X_\lambda$ is locally compact, and every neighborhood of X in R^p contains some X_λ; also, every neighborhood of $A = \bigcap_{\lambda \in \Lambda} A_\lambda$ in R^p contains some A_λ. Then</u>

$$\Omega_\sigma^j(X, A) \cong \varinjlim \{\Omega_\sigma^j(X_\lambda, A_\lambda)\}_{\lambda \in \Lambda} .$$

Proof. We have to show (cf. $[D_1]$,VIII,5.18) that

(i) every $x \in \Omega(X,A)$ comes from some $\Omega(X_\lambda,A_\lambda)$, and

(ii) if $x_\lambda \in \Omega(X_\lambda,A_\lambda)$ is such that $x_\lambda|(X,A) = 0$ then $x_\lambda|(X_\mu,A_\mu) = 0$ for some $\mu > \lambda$. In order to prove (ii), we represent x_λ by some proper $g_\lambda : M \to V_\lambda$ with $g_\lambda(M) \subset V_\lambda - A_\lambda$; then $x_\lambda\ |(X,A) = 0$ implies that X has an open neighborhood V in V_λ such that $g_\lambda^{-1}V \to V$ bounds some proper $G : W \to V$ with $G(W) \subset (V - A)$. Now V contains some X_{λ_1} and $V - G(W)$ contains some A_{λ_2} ; by directedness there exists $\mu > \lambda,\lambda_1,\lambda_2$, hence $X_\mu \subset V$, $A_\mu \subset (V - G(W))$. It follows that $g_\lambda^{-1} = \partial W$ represents the zero element of $\Omega(X_\mu,A_\mu)$, i.e. $x_\lambda|(X_\mu,A_\mu) = 0$. - The proof of (i) is even simpler. ∎

(2.14) Induced homomorphisms $\alpha^*: \Omega X \to \Omega Y$.

So far we've only dealt with inclusion maps $Y \subset X$, now we consider arbitrary continuous maps $\alpha : Y \to X$, where $X \subset R^p$, $Y \subset R^q$ are locally closed. As before, we first treat the case where X,Y are open in R^p,R^q . Let $A \subset X$, $B \subset Y$ relatively closed subsets such that $\alpha(B) \subset A$. Elements $a \in \Omega_\sigma^j(X,A)$ are represented by proper σ-structured manifolds $g : M \to X$ with $g(M) \subset (X - A)$, $\dim M = p - j$. Suppose first that g is smooth, α is smooth in a neighborhood of $\alpha^{-1}(g(M))$, and (g,α) are transverse. The latter means (cf. 1.12) that $\alpha \times g : Y \times M \longrightarrow X \times X$ is transverse to the diagonal submanifold of $X \times X$; in our case where X is an open subset of R^p it simply means that $0 \in R^p$ is a regular value of the difference mapping

(2.15) $\langle\alpha,g\rangle : Y \times M \to R^p$, $\langle\alpha,g\rangle\ (y,m) = \alpha(y) - g(m)$.

It follows that $\langle\alpha,g\rangle^{-1}\ (0)$ - which coincides with the pullback $Y \times_X M$ of $Y \xrightarrow{\alpha} X \xleftarrow{g} M$ - is a smooth submanifold of $Y \times M \subset R^q \times M$ with trivialized (by $T\langle\alpha,g\rangle = (T\alpha,-Tg)$) tubular neighborhood (~ normal bundle); thus $R^p \times (Y \times_X M)$ is represented as an open subset of $R^q \times M$. It is therefore σ-structured and hence $Y \times_X M$ is σ-structured. Since pullbacks of proper maps are proper, $g' : Y \times_X M \to Y$ is a proper σ-manifold over Y . Clearly, $g'(Y \times_X M) \subset (Y - \alpha^{-1}A) \subset Y - B$. By definition, g' represents $\alpha^*(a) \in \Omega_\sigma^j(Y,B)$, up to a sign $(-1)^{p+qp}$; thus

$$(2.16) \qquad \alpha^* : \Omega_\sigma^j(X,A) \longrightarrow \Omega_\sigma^j(Y,B) ,$$

$$\alpha^*[g:M \to X] = (-1)^{p+qp}[Y \times_X M \to Y]$$

if α , g are smooth (where it matters; cf. above) and transverse.

The sign $(-1)^{p+pq}$ reflects the behavior of σ-structures under the linear maps $R^p \times R^q \approx R^q \times R^p$ and $-\mathrm{id}(R^p)$. For many purposes it can be ignored, e.g. for determining (co-)kernels etc. For a better understanding of the sign the reader might consider the examples 2.17, and the proof of functoriality (2.18).

If α, g are (smooth but) not transverse, i.e. if $\langle\alpha,g\rangle$ is smooth in a neighborhood of $\langle\alpha,g\rangle^{-1}(0)$ but 0 is not a regular value then one is tempted to replace 0 by a nearby value z which is regular - but then $\langle\alpha,g\rangle^{-1}(z)$ may fail to be proper over Y . However, standard transversality arguments (cf. [H], Prop.3.2.7; [A-R], §4) show that one can approximate α by a homotopic map $\alpha' = \alpha-\xi : Y \to X$, where $\xi : Y \to R^p$ is a small correction function which vanishes on a neighborhood of B and such that (α',g) are transverse. Then $\alpha^*[g] = \alpha'^*[g] = [\langle\alpha',g\rangle^{-1}(0) \to Y]$, by definition. If α or g are not even smooth then one approximates them by homotopic smooth mappings and proceeds as above; for α one only requires smoothness near $\alpha^{-1}(gM)$, and one does not move it near B . Applying the same arguments to proper σ-manifolds with boundary $W \to X$ shows that $\alpha^*(a)$ does not depend on the choice of the representative $[g]$. Replacing Y by $[0,1] \times Y$

and using the same arguments shows that $\alpha^*(a)$ does not depend on the choice of the approximation α' ; moreover, it shows that α^* depends only on the homotopy class of α (in fact, only on the (co)bordism class of α - in a certain sense which we shall not discuss here; compare 4.12). Thus, $\alpha^* : \Omega_\sigma^j(X,A) \longrightarrow \Omega_\sigma^j(Y,B)$ is defined now for arbitrary continuous maps $\alpha : (Y,B) \longrightarrow (X,A)$ such that $X \subset R^p$, $Y \subset R^q$ are open. Moreover, $\alpha_o \simeq \alpha_1 \Rightarrow \alpha_o^* = \alpha_1^*$ (homotopy invariance).

Before we proceed to more general X,Y we consider some

(2.17) Examples. If $X = pt$ is a single point then $\Omega_\sigma^j X$ is the bordism group of compact σ-manifolds N of dimension $-j$. For general open $X \subset R^p$ again the map $X \to pt$ induces

$\Omega_\sigma^j(pt) \longrightarrow \Omega_\sigma^j X$, $[N] \longmapsto [proj : X \times N \to X]$.

Any $\alpha : Y \to X$ must satisfy $\alpha^*[X\times N\to X] = [Y\times N\to Y]$ — if we are to have functoriality. We check this equality with our definition 2.16. Assuming α smooth we consider

$$\langle\alpha,g\rangle : Y \times X \times N \longrightarrow R^p , \langle\alpha,g\rangle(y,x,n) = \alpha(y) - x .$$

Thus $\langle\alpha,g\rangle^{-1}(0) = \{(y,x,n) | x = \alpha(y)\} = \text{graph}(\alpha) \times N \approx Y \times N$.

The trivialization of the normal bundle of $\langle\alpha,g\rangle^{-1}(0) = Y \times N$ is given by

$$\tau : R^p \times Y \times N \to R^q \times R^p \times N , \tau(x,y,n) = (y,\alpha(y)-x,n) .$$

The σ-structure on $Y \times N$ which is induced by τ from the standard structure of $R^q \times R^p \times N$ is therefore equal to the standard structure multiplied by the effect of $(x,y) \to (y,-x)$, i.e. by $(-1)^{p+qp}$ - as required.

For another example consider an inclusion map $\alpha : Y \subset X$ of open subsets of R^p , thus $\alpha(y) = y$. Then α is transverse to every smooth $g : M \to X$ and $\langle\alpha,g\rangle^{-1}(0) = \{(y,m) | y = g(m)\} \approx g^{-1}Y$. The trivialization of the normal bundle of $g^{-1}Y$ is given by

$$\tau : R^p \times g^{-1}Y \longrightarrow R^p \times M \ , \ \tau(x,m) = (g(m)+x,m) \ .$$

This induces the same structure as the inclusion $(x,m) \mapsto (x,m)$, i.e.

$$\alpha^*[g : M \to X] = [g^{-1}Y \to Y] \ .$$

In other words, α^* coincides with ι^* as defined in (2.3) resp. 2.4; note that $(-1)^{p+pq} = +1$ if $p = q$.

(2.18) Proposition (functoriality). If $X \subset R^p$, $Y \subset R^q$, $Z \subset R^s$ are open sets and $Z \xrightarrow{\beta} Y \xrightarrow{\alpha} X$ are continuous maps then $(\alpha\beta)^* = \beta^*\alpha^*$; $\Omega^j_\sigma X \longrightarrow \Omega^j_\sigma Z$; similarly, for relative groups. The identity map $id : X \to X$ induces the identity endomorphism of $\Omega^j_\sigma(X,A)$.

Proof. After smoothing the maps and making them transverse where appropriate we have to consider a diagram

$$\begin{array}{ccccc} L & \xrightarrow{\beta} & N & \xrightarrow{\alpha'} & M \\ \downarrow{g''} & & \downarrow{g'} & & \downarrow{g} \\ Z & \xrightarrow[\beta]{} & Y & \xrightarrow[\alpha]{} & X \end{array} \qquad (2.19)$$

where both squares are pullback diagrams with transverse (α,g) and (β,g') ; the outer square is then also pullback with transverse $(\alpha\beta,g)$. X,Y,Z are euclidean open sets (cf. above) and M (hence N , hence L) are σ-structured. It is clear that L equals $Z \times_Y N = Z \times_Y (Y \times_X M)$ as well as $Z \times_X M$ so that $\beta^*\alpha^*[g] = (\alpha\beta)^*[g]$ except (perhaps) for the σ-structure on L . The σ-structure of $L = Z \times_Y N = Z \times_Y (Y \times_X M)$ is obtained by considering the embeddings

(2.20) $L = Z \times_Y N = \langle \beta, g' \rangle^{-1}(0) \overset{q}{\subset} Z \times N = Z \times \langle \alpha, g \rangle^{-1}(0) \overset{p}{\subset} Z \times Y \times M$

with trivialized normal bundles; the superscripts q,p indicate the fibre $\mathbb{R}^q$, $\mathbb{R}^p$ of these normal bundles. The composite embedding has trivialized $(p+q)$-normal bundle; the trivialization is given by the derivate at $L = Z \times_Y Y \times_X M$ of the following map

(2.21) $Z \times Y \times M \longrightarrow \mathbb{R}^p \times \mathbb{R}^q$, $(z,y,m) \longmapsto (\alpha(y)-g(m), \beta(z)-y)$.

The σ-structure of $L = Z \times_X M$ is obtained by considering the embedding $L = Z \times_X M = \langle \alpha\beta, g \rangle^{-1}(0) \overset{p}{\subset} Z \times M$; to make it comparable with (2.20) we compose with the graph-embedding

$$(1,\beta) \times 1 \ : \ Z \times M \overset{q}{\subset} Z \times Y \times M$$

with obviously trivialized normal bundles. The composite

$L = Z \times_X M = \langle \alpha\beta, g \rangle^{-1}(0) \overset{p}{\subset} Z \times M \overset{q}{\subset} Z \times Y \times M$

has its normal bundle trivialized by the derivative at L of the following map

(2.22) $Z \times Y \times M \longrightarrow \mathbb{R}^p \times \mathbb{R}^q$, $(z,y,m) \longmapsto (\alpha\beta(z)-g(m), y-\beta(z))$.

The two maps (2.21),(2.22) have opposite second components; their first components (agree at L and they) have the same derivate at L (in order to calculate the derivative transverse to L write $y = \beta(z)+a$; the derivatives of $z \mapsto \alpha\beta(z)$ and $z \mapsto \alpha(\beta(z)+a)$ are the same at $a = 0$) . Therefore the trivializations induced by (2.21), (2.22) agree up to $-\mathrm{id}(\mathbb{R}^q)$. When we write down the maps $L \times \mathbb{R}^p \times \mathbb{R}^q \overset{o}{\subset} Z \times Y \times M$ which represent the trivialized tubular neighborhood ($\simeq$ normal bundle) we also have to exchange some factors (like $Z \times \mathbb{R}^p \sim \mathbb{R}^p \times Z$) which produces further signs on the σ-structures; altogether one finds a sign $(-1)^{q+pr+pq+qr}$, by which the σ-structured of $L = Z \times_Y N$ and $L = Z \times_X M$ differ. But this is also the difference (or quotient, rather) of the signs $(-1)^{p+pq}.(-1)^{(q+qr)}$ and $(-1)^{p+pr}$ which we introduced in 2.16, hence $(\alpha\beta)^*[g] = \beta^*\alpha^*[g]$. In fact the explicit calculation of signs can be avoided: Once it is clear that $(\alpha\beta)^*$, $\beta^*\alpha^*$ differ by a sign which depends only on p,q,r we can check on examples that the sign is $+1$. Such examples were discussed in 2.17; we can even take the example $M = X$, $g = id$.

The second example of 2.17 also shows $(id)^* = id$. ∎

<u>(2.23) Induced homomorphisms $\varphi^* : \Omega X \to \Omega X$; continued</u>. Mappings $\varphi : Y \to X$ between locally compact subsets of euclidean spaces can always be extended to open neighborhoods, and uniquely so up to homotopy (in some smaller neighborhoods). This allows to extend homotopy invariant functors of (continuous maps between) open euclidean sets to locally compact euclidean sets. This is a standard procedure. It is discussed in detail (for another example) in [D_1] VIII, 6. It should therefore be enough if we just indicate some steps here, without proofs.

Given $Y \subset R^q$ locally compact, let $V \subset R^q$ an open neighborhood of V such that Y is relatively closed in V. If $X \subset R^p$ is also locally compact and $\varphi : Y \to X$ is continuous then (by Tieze's lemma) φ extends to some $\Phi : Y \to R^p$. For every open neighborhood $U \subset R^p$ of X the map $\Phi_U = \Phi|\Phi^{-1}U : \Phi^{-1}U \to U$ induces

$$(2.24) \quad \Phi_U^* : \Omega_\sigma^j U \longrightarrow \Omega_\sigma^j(\Phi^{-1}U) \longrightarrow \Omega_\sigma^j Y .$$

This is a direct system of homomorphism, indexed by $\{U\}$. Passing to $\varinjlim$ it induces

$$(2.25) \quad \varphi^* = \{\Phi_U^*\} : \Omega_\sigma^j X \longrightarrow \Omega_\sigma^j Y .$$

<u>Explicitely</u>, elements $x \in \Omega_\sigma^j X$ are represented by σ-structured

proper manifolds $g : M \to U$; deform $\Phi_U : \Phi^{-1}U \to U$ to make it (smooth and) transverse to g , then $(\Phi^{-1}U) \times_U M \longrightarrow \Phi^{-1}U$ is σ-structured (as under 2.14, using $\langle \Phi_U, g \rangle$) and proper ; it represents the element $(-1)^{p+pq}\, \varphi_*(x) \in \Omega_\sigma^j\, Y$, by definition of $\varphi_*(x)$. If $A \subset X$, $B \subset Y$ are closed subsets such that $\varphi(B) \subset A$, and $g : M \to U$ represents $a \in \Omega_\sigma^j(X,A)$ then $g(M) \subset (X-A)$; it suffices then to make Φ_U smooth and transverse to g in a neighborhood of $\Phi_U^{-1}(gM)$, without changing it near B . Again $(\Phi^{-1}U) \times_U M \longrightarrow \Phi^{-1}U$ will then represent the element $(-1)^{p+pq}\varphi_*(a) \in \Omega_\sigma^j(Y,B)$.

This now makes Ω_σ^j a contravariant functor on maps $\varphi : (Y,B) \longrightarrow (X,A)$, as above. Together with the preceding results it shows that

<u>2.26 Theorem</u>. $\Omega_\sigma^* = \{\Omega_\sigma^j\}_{j \in \mathbb{Z}}$ <u>is a cohomology theory on the category of continuous maps between pairs (X,A) , where X is locally compact in some R^p , and A is relatively closed in X . The theory is continuous (commutes with $\varinjlim$ in the sense of 2.13), and it is strongly additive</u>, $(\Omega(\bigoplus_{i=0}^{\infty} X_i) = \prod_{i=0}^{\infty} (\Omega X_i))$. ∎

We claim, of course that <u>Ω_σ^* is dual to the (rather obvious) bordism homology theory Ω_*^σ</u> . This is essentially the following

<u>2.27 Theorem</u>. <u>If $A \subset X$ are compact subsets of R^p then</u> $\Omega_\sigma^j(X,A) \cong \Omega_{p-j}^\sigma(R^p-A, R^p-X)$. <u>More generally, the same isomorphisms hold if $A \subset X$ are closed in R^p and $X - A$ is bounded</u>.

<u>Proof</u>. Assume $A = \emptyset$, the general case being similar. Elements $x \in \Omega_\sigma^j X$ are represented by proper σ-manifolds $g : M \to V$, where V is an open neighborhood of X in R^p , and $\dim(M) = p - j$. Choose a smooth function $\tau : M \to [0,1]$ with compact carrier

$(=\overline{\tau^{-1}(0,1]})$ such that $\tau|g^{-1}X = 1$. Every regular value $\varepsilon \in (0,1)$ of τ then yields a compact σ-manifold with boundary $N = \tau^{-1}[\varepsilon,1]$, $\partial N = \tau^{-1}(\varepsilon)$, and a map $g_\varepsilon = g|N : (N,\partial N) \longrightarrow (R^p, R^p - X)$ whose bordism class $y = [g_\varepsilon] \in \Omega^\sigma_{p-j}(R^p, R^p-X)$ depends only on x.

Conversely, elements $y \in \Omega^\sigma_{p-j}(R^p,R^p-X)$ are represented by continuous maps $f : (N,\partial N) \longrightarrow (R^p,R^p-X)$, where N is a compact σ-manifold of dimension $p - j$. Let $V = R^p - f(\partial N)$. This is an open neighborhood of X, and $g = f|M : M = N - \partial N \longrightarrow V$ is a proper σ-manifold whose cobordism class $x = [g] \in \Omega^j_\sigma X$ depends only on y. Moreover, $x \mapsto y$, $y \mapsto x$ are inverse to each other. ∎

(2.28) $\Omega^*_\sigma(X,A)$ for non-closed A. This is of considerable interest (in particular for open $A \subset X$; compare 2.3) but we ignored it so far in order to keep the presentation simple. We give the definition now but omit further details and verifications.

Let $A \subset X$ locally compact subsets of some R^p. Since A is locally closed in X it has an open neighborhood U in X in which it is closed. Now U is defined by a continuous function $d : X \to [0,+\infty)$; thus, $U = d^{-1}(0,+\infty)$. The set

$$U^\natural = \{(t,x) \subset R \times X \mid t\cdot d(x) \geq 1\}$$

is closed in $R \times X$, and homeomorphic with $[0,+\infty) \times U$ under the map

$$U^\natural \longrightarrow [0,+\infty) \times U , \quad (t,x) \mapsto (t - 1/d(x) , x) .$$

The set

$$A^\natural = \{(t,x) \in R \times X \mid x \in A , \text{ and } t\cdot d(x) \geq 1\} = U^\natural \cap (R \times A)$$

is closed in $U^{\natural}$ and hence in $\mathbb{R} \times X$: it is homeomorphic with $[0,+\infty) \times A$. We now <u>define</u>

$$(2.29) \qquad \Omega^j(X,A) = \Omega^j(\mathbb{R} \times X , A^{\natural}) .$$

This does not depend on the choice of d . (If d' is a second choice then $(\mathbb{R}\times X, A^{\natural}) \approx (\mathbb{R}\times X, A^{\natural'})$ - one can choose this hoemomorphism of the form $(t,x) \mapsto (\eta(t,x),x)$, with $\eta(t,x) = t$ for $x \notin U$ or $t+1 < \min(1/d(x), 1/d'(x))$). Also, if A is closed in X then the new definition 2.29 agrees with the old one (2.7,2.8), as the projection $(\mathbb{R}\times X, A\ ) \longrightarrow (X,A)$ shows.

<u>2.30 Exercise</u>. If $A \subset X$ are open subsets of $\mathbb{R}^p$ such that $(X-A)$ is a neighborhood retract then $\Omega^j_\sigma(X,A)$ can be defined <u>as if A were closed in X</u> (cf.2.1), i.e. by representatives $g : M$ $g : M \to X$ which are proper, σ-structured, $(p-j)$-dimensional and satisfy $g(M) \subset (X-A)$, and by cobordism $G : W \to X$ of the same kind.

(2.31) <u>Steenrod bordism groups $\hat{\Omega}^\sigma_j$</u> . Ordinary bordism groups Ω^σ_j (in contrast to cobordism groups Ω^j_σ) behave rather badly on closed (compact) subsets of $\mathbb{R}^p$ which are not neighborhood retracts (example: $\overline{\text{graph}(x \mapsto \sin(1/x))}$. The difficulty is the same as for singular homology H_j ; there it is resolved by using Steenrod homology. In a recent paper by Kahn-Kaminker-Schochet [KKS] this solution was generalized to arbitrary (co-)homology theories. The result is very satisfactory but the main definition resp. the existence proof in [KKS] is somewhat technical and complicated. In our case, however, there is a simple geometric interpeetation for the (generalized) Steenrod groups, as follows.

If $A \subset K$ are closed subsets of $\mathbb{R}^p$ such that $\overline{K-A}$ is compact their Steenrod groups are given (cf. [KKS],theorem B) by

(2.32) $\hat{\Omega}^{\sigma}_{j}(K,A) \cong \Omega^{p-j}_{\sigma}(R^p-A,R^p-K) = \Omega^{p-j}_{\sigma}(R\times(R^p-A),(R-K)^{\natural})$,

the last equality by 2.29. Let us assume $A = \emptyset$, the general case being very similar. Thus K is compact in R^p, and $\hat{\Omega}^{\sigma}_{j}K \cong \Omega^{p-j}_{\sigma}(R\times R^p,(R^p-K)^{\natural})$, where $(R^p-K)^{\natural} = \{(t,x) \in R\times R^p \mid t\cdot d(x) \geq 1\}$, $d(x)$ = distance (x,K).

The complement of this set,

$$\hat{K} = R \times R^p - (R^p-K)^{\natural} = \{t,x) \in R\times R^p \mid t\cdot d(x) < 1\}$$

is an open neighborhood of $R\times K$ in $R\times R^p$ whose section with $t = \text{const.} \geq 0$ is the $\frac{1}{t}$ - neighborhood of K in R^p; it shrinks down to K as $t \to +\infty$. The elements $\xi \in \hat{\Omega}^{\sigma}_{j}K \approx \Omega^{p-j}_{\sigma}(R\times R^p, R\times R^p-\hat{K})$ are represented by $(j+1)$-dimensional σ-manifolds $\gamma : N \to \hat{K}$ such that the composition $N \to \hat{K} \subset R\times R^p$ is proper; we can assume that γ is smooth. The composition

$$\pi : N \xrightarrow{\gamma} \hat{K} \subset R \times R^p \xrightarrow{\text{proj}} R$$

is a smooth function on N; almost all $t \in R$ are regular values of π. For these $t > 0$ the <u>regular section</u> $N_t = \pi^{-1}(t)$ is a j-dimensional compact σ-structured manifold over the $1/t$ - neighborhood $\hat{K}_t$ of K. As t grows $N_t \to \hat{K}_t$ moves closer and closer to K; moreover, if $t' > t > 0$ are two regular values then $\pi^{-1}[t,t']$ is a bordism between N_t and $N_{t'}$, which takes place in the $1/t$ - neighborhood $\hat{K}_t$ of K. This then is <u>the geometric meaning of an element $\xi \in \hat{\Omega}^{\sigma}_{j}K$</u>, namely <u>a variable j-dimensional compact σ-manifold $N_t \to R^p$ closing in on K as $t \to +\infty$</u>.

If K happens to be a <u>neighborhood retract in R^p</u> (an ENR) then we can retract the $1/t$-neighborhood K_t onto K for some t; say $r : K_{t_o} \to K$, $r|K = id$. For regular $t \geq t_o$ the composite

$N_t \to K_t \xrightarrow{r} K$ then represents a well-defined element $\xi' = [N_t \to K] \in \Omega_j^\sigma K$, and the mapping $\xi \longmapsto \xi'$ is an isomorphism. Similarly for pairs (K,A) , i.e.

(2.33) Proposition. If $A \subset K$ are closed subsets of R^p such that $\overline{K-A}$ is compact and if both A and K are neighborhood retracts (ENR) then

$$\Omega_j^\sigma(K,A) \cong \Omega_\sigma^{p-j}(R^p-A, R^p-K) \ . \ \blacksquare$$

3. Cobordism groups of manifolds; duality.

(3.1) In this section we assume that the structure functor σ is multiplicative (cf. 1.16) - although some parts could do with weaker ad hoc assumption. Multiplicativity implies, in particular, a canonical σ-structure on euclidean open sets or, more generally, on (stably) parallelized manifolds. Also, the σ-structures on the tangent bundle TX of a manifold X are in canonical bijective correspondance with the σ-structures on the normal bundle $T'X$ (cf. 1.22).

3.2 Cobordism groups of manifolds. Let X an n-dimensional smooth submanifold of $\mathbb{R}^p$ with (open) tubular neighborhood $T'X$ (= normal bundle); let $X \xrightarrow{i} T'X \xrightarrow{r} X$ inclusion and retraction, $ri = id$, $ir \simeq id$. Then $\Omega_\sigma^j(X) \cong \Omega^j(T'X)$, via i^* resp. r^*. Thus, all elements of $\Omega_\sigma^j X$ are represented by proper $(p-j)$-dimensional σ-manifolds $g : M \to T'X$, and two such g_1, g_2 represent the same element of $\Omega_\sigma^j X$ iff they are cobordant over $T'X$. Moreover, we can assume that the representatives g as well as the cobordisms are smooth and transverse to i. Then $N = g^{-1}X = X \times_{T'X} M$ is an $(n-j)$-dimensional proper smooth manifold over X, say $h : N \to X$; similarly for cobordisms, i.e. proper manifolds with boundary. The original $g : M \to T'X$ can be recaptured from $h : N \to X$, up to cobordism, because $[g : M \to T'X] = (ir)^*[g : M \to T'X]$, and the latter is represented by $T'X_{ir}\times_g M = T'X_r\times_h N = T'X \times_X N$. Of course, N may not be σ-structured (not even structurable). However, the map $h : N \to X$ is structured - essentially by definition (1.10) because σ-structures of h are σ-structures of $TN \oplus h^*T'X = T(T'X \times_X N)|N$, and σ-structures of the bundle $T(T'X \times_X N)$ are σ-structures of the manifold $T'X \times_X N$ by 1.7.

Conversely, if $h : N \to X$ is a proper σ-structured smooth map with $\dim N = n-j$ then $M = (T'X) \times_X N$ is a $(p-j)$-dimensional σ-structured (as above, using 1.10) smooth manifold with proper projection $g : M = (T'X) \times_X N \longrightarrow T'X$, and $g^{-1}X = N$, $g|g^{-1}X = h$.

These considerations lead to the following (auxiliary) definition (3.3) and proposition (3.4).

(3.3) Definition. For X a smooth manifold (in some R^p ; $\partial X = \emptyset$) consider all σ-structured proper maps $h : N \to X$, where N is a smooth manifold with $\partial N = \emptyset$. Two such $h_o : N_o \to X$, $h_1 : N_1 \to X$ are said to be cobordant, $h_o \sim h_1$, if a proper σ-structured map $H : W \to X$ exists such that $\partial W = N_o \oplus N_1$, $H|\partial W = (h_o, h_1^-)$. In other words, we proceed exactly as in 2.1 except that X now is an arbitrary smooth manifold (in some R^p, but not open in general), and it is the map h which is structured now, not the domain N of the map. We obtain groups $\Lambda_\sigma^j X$ in this way; the elements of $\Lambda_\sigma^j X$ are cobordism classes $[h : N \to X]$ of maps as above with $\dim N = \dim X - j$.

Similarly, if A is a closed subset of X we obtain $\Lambda_\sigma^j(X,A)$ by requiring in addition that $h(N) \subset (X-A)$. The formal properties of these groups are similar, and similarly proved, as for Ω_σ^j. In fact, the preceding considerations under 3.2 show that

(3.4) Proposition. $\Lambda_\sigma^j(X,A) \cong \Omega_\sigma^j(X,A)$,
via $[N \to X] \longmapsto [(T'X) \times_X N \longrightarrow T'X]$. ∎

More generally, the definition of $\Lambda_\sigma^j(X,A)$ (and the proposition 3.4) can be extended to locally closed A, as in 2.28.

(3.5) Examples. (i) For every manifold X as above the identity map $1_X : X \to X$ is canonically σ-structured (1.11); it represents a canonical element in $\Lambda_\sigma^o X = \Omega_\sigma^o X$ which we still denote by $1 = 1_X$.

If $X \neq \emptyset$ then $1_X \neq 0$ (exercise).

(ii) If $j > \dim X$ then $\Omega_\sigma^j(X,A) = \Lambda_\sigma^j(X,A) = 0$.

(iii) For $j = \dim X = n$, elements of $\Omega_\sigma^n X = \Lambda_\sigma^n X$ are represented by σ-structured maps $g : N \to X$ with $\dim N = 0$, hence N discrete. By additivity, it suffices to calculate $\Omega_\sigma^n X$ for connected X . If X is not compact then any point, $pt \to X$, is the boundary of a proper map $[0,+\infty) \to X$, hence $\Omega_\sigma^n = 0$. If X is compact N must be finite and g is cobordant to a constant map; it easily follows that $\Omega_\sigma^n X = \Omega_\sigma^0(pt)$ if X is orientable, and $= \Omega_\sigma^0(pt) \otimes \mathbb{Z}/2\mathbb{Z}$ otherwise. For $\Omega_\sigma^0(pt)$ we have:

(iv) Let $1^- \in \sigma(\mathbb{R}^0)$ be the opposite structure of 1 (cf.1.6 and 1.19). If $1^- = 1$ then $\Omega_\sigma^0(pt)$ is the (additive group of the) mod 2 group ring of $\sigma(\mathbb{R}^0)$. If $1^- \neq 1$ then $\Omega_\sigma^0(pt)$ is the integral group ring of the quotient group $\sigma(\mathbb{R}^0)/_{\{1,-1\}}$.

<u>3.6 Theorem (duality ; compare 2.27). Let X a σ-structured manifold of dimension n , and let $A \subset B$ be closed subsets of X such that $\overline{B-A}$ is compact. Then</u>

$$\Omega_\sigma^j(B,A) \cong \Omega_{n-j}^\sigma(X-A, X-B) .$$

<u>Proof</u>. We shall only prove the two special cases (1) $B = X$, and (2) $A = \emptyset$; these two cases exhibit the essential features, and the general case reduces to it in a standard way (compare $[D_1]$,VIII,7). As for (1), elements of $\Omega_\sigma^j(X,A) = \Lambda_\sigma^j(X,A)$ are represented by σ-structured proper maps $h : N \to X$ such that $h(N) \subset (X-A)$, $\dim N = n-j$. Since $\overline{X-A}$ is compact and $N = h^{-1}(\overline{X-A})$ we see that h is proper iff N is compact. The representatives of elements in $\Lambda_\sigma^j(X,A)$ are therefore simply σ-structured maps $N \to (X-A)$ of compact manifolds N . But these are also the representatives for bordism $\Omega_{n-j}^\sigma(X-A)$ - except that for Ω_{n-j}^σ the domain N has to be σ-structured. But the set of σ-structures $\sigma(h) = \sigma(TN \oplus h^*T'X)$ is

in bijective correspondence with the set of σ-structures $\sigma(TN)$ because X is σ-structured, i.e. $T'X$ is, and hence $g^*T'X$; the correspondence $\sigma(TN) \xrightarrow{\approx} \sigma(h)$ is obtained by multiplying $u \in \sigma(TN)$ with the given (fixed) element $u_o \in \sigma\,(h^*T'X)$.

For case (2) we essentially copy the argument for 2.27: Since

$$\Omega^j_\sigma B = \varinjlim \{\Omega^j_\sigma V \mid V \text{ open nbhd of } B \text{ in } X\}$$

by (2.13), and $\Omega^j_\sigma V = \Lambda^j_\sigma$, we can represent the elements $x \in \Omega^j_\sigma B$ by σ-structured proper maps $h : N \to V$, with $\dim N = n-j$, $\partial N = \emptyset$. As for (1), structuring the map h is equivalent to structuring its domain N . Choose a smooth function $\tau : N \to [0,1]$ with compact carrier $\overline{\tau^{-1}(0,1]}$ such that $\tau \mid h^{-1}B = 1$. Every regular value $\varepsilon \in (0,1)$ of τ then yields a compact σ-manifold $L = \tau^{-1}[\varepsilon,1]$ and a map $h_\varepsilon : (L,\partial L) \longrightarrow (X,X-B)$, hence an element $y = [h_\varepsilon] \in \Omega^\sigma_{n-j}(X,X-B)$. This is the image of x . The inverse map $\Omega^\sigma_{n-j}(X,X-B) \longrightarrow \Omega^j_\sigma B$, and further details are as in the proof of 2.27. ∎

<u>3.7 Exercise (compare 2.33).</u> If $A \subset B \subset X$ are as in 3.6 then

$$\hat\Omega^\sigma_j(B,A) \cong \Omega^{n-j}_\sigma(X-A,X-B)$$

where $\hat\Omega$ denotes Steenrod bordism (2.31). If, in addition, both A and B are neighborhood retracts then

$$\Omega^\sigma_j(B,A) \cong \Omega^{n-j}_\sigma(X-A,X-B) \ .$$

<u>(3.8) Gysin homomorphism</u>. If $\alpha : X \to Y$ is a proper σ-structured map between smooth manifolds X,Y of dimension m,n then the <u>Gysin homomorphism</u> $\alpha_!$ is defined as follows

(3.9) $\alpha_! : \Omega^j_\sigma X \longrightarrow \Omega^{j+n-m}_\sigma Y \ , \quad \alpha_![g : M \to X] = [\alpha g : M \to Y]$,

where g is proper and σ-structured, αg is equipped with the composite σ-structure (cf. 1.24), $\dim M = m-j$.

More generally, if $\alpha : X \to Y$ is σ-structured but not necessarily proper we can still define

$$(3.9') \qquad \alpha_! : \Omega_\sigma^j(X,A) \longrightarrow \Omega_\sigma^{j+n-m}(Y,B)$$

(by the same formula 3.9) for all closed subsets $A \subset X$, $B \subset Y$ such such that $\alpha \mid \overline{X-A}$ is proper, and $A \supset \alpha^{-1}B$.

Clearly, $(\beta\alpha)_! = \beta_!\alpha_!$ whenever this makes sense, and $\mathrm{id}_! = \mathrm{id}$, where $\mathrm{id} = \mathrm{id}_X$ on the left is to be taken with the canonical structure (cf. 1.24). In other words, the groups $\Lambda_j X = \Omega^{\dim X-j}\, X$ (and their relative generalizations) are covariant functors on the category of (proper) σ-structured maps (compare this with $\mathfrak{C}_\sigma$ under 1.24).

(3.10) Proposition. If

$$\begin{array}{ccc} X' & \xrightarrow{\beta'} & X \\ \Big\downarrow{\scriptstyle \alpha'} & & \Big\downarrow{\scriptstyle \alpha} \\ Y & \xrightarrow{\beta} & Y \end{array}$$

is a pullback-diagram $(X' = Y' \times_Y X)$, where (α,β) are smooth and transverse, α is σ-structured and proper, and α' has the induced structure (cf. 1.12) then $\beta^*\alpha_! = \pm\, \alpha'_!\beta'^*$. Similarly in the relative case. The proof is easy with representatives (cf.the lines just before 4.16). The sign $\pm$ is left open because I didn't make the signs precise in 1.12 (cf. 1.15). A proper choice of signs (as in 2.16,2.18) will lead to $+1$ here.

(3.11) Example (Thom-Gysin-homomorphism. Let $\pi : E \to X$ a (k-plane) vector bundle over the manifold X , with zero-section $\iota : X \to E$. If π is σ-structured then there is a unique way to σ-structure ι such that $\pi\iota$ (and $\iota\pi$) has the canonical identity structure. This follows from 1.20 and 1.24. Actually, $\sigma(\iota) = \sigma(\pi')$

as pointed out in 1.11, so that the result also follows from 1.22. The map ι is proper so that we have the Gysin-homomorphism

$$(3.12) \qquad \iota_! \; : \; \Omega_\sigma^j X \longrightarrow \Omega_\sigma^{j+k}(E, E-U)$$

where U is any neighborhood of the zero-section. The map π is not proper but if U is contained in a tube U^ρ of finite radius ρ then $\pi \mid \overline{U}$ is proper and we can define

$$(3.13) \qquad \pi_! \; : \; \Omega_\sigma^{j+k}(E, E-U) \longrightarrow \Omega_\sigma^j X \; .$$

Since $\pi\iota = \mathrm{id}$ we have $\pi_! \iota_! = \mathrm{id}$. The other composite, $\iota\pi$, is homotopic to id_E but we have to make sure that the homotopy moves U within U. This is alright if $U = U^\rho$ is itself a tube, thus

$$(3.14) \qquad \Omega_\sigma^{j+k}(E, E-U^\rho) \cong \Omega_\sigma^j X \qquad \text{(\underline{Thom isomorphism})} \; .$$

If we allow open subsets for relative groups, as in 2.28, 3.14 becomes

$$(3.14') \qquad \Omega_\sigma^{j+k}(E, E-\iota X) \cong \Omega_\sigma^j X \; .$$

The geometry of the isomorphism $\iota_!$ is extremely simple: If $g : M \to X$ represents $x \in \Lambda_\sigma^j X \cong \Omega_\sigma^j X$ then $g : M \to E$ represents $(\iota_! x) \in \Omega_\sigma^{j+k}(E, E-U)$. - Similar arguments will establish <u>relative Thom isomorphisms</u>

$$(3.15) \qquad \Omega_\sigma^j(X,A) \cong \Omega_\sigma^{j+k}(E, (E-U^\rho) \cup E_A) \; , \text{ where } E_A = \pi^{-1}A \; .$$

We can replace X by any open subset $V \subset X$ and obtain $\Omega_\sigma^j V \cong \Omega_\sigma^{j+k}(E_V, E_V - U_V^\rho)$, where $E_V = \pi^{-1} V$, etc. If Z is any locally compact subset of X then we can let V range over all open neighborhoods of Z (or a cofinal subsystem), and pass to $\varinjlim$. We obtain

$$(3.16) \qquad \Omega_\sigma^j Z \cong \Omega_\sigma^{j+k}(E_Z, E_Z - U_Z^\rho) \; .$$

Since every vector-bundle over Z extends to a neighborhood this provides Thom-isomorphisms for σ-structured vector-bundles over non-manifolds (cf. also the remark 1.5 (iv)) .

4. Products in (co-)bordism.

These are fairly obvious. We shall therefore confine ourselves to describing the definitions and to commenting on some noteworthy aspects. - As in section 3, σ is assumed to be a multiplicative structure functor.

(4.1) Cross- and Cup-products. If $X \subset R^p$, $Y \subset R^q$ are locally compact we define exterior- or cross-products

$$(\Omega_\sigma^j X) \times (\Omega_\sigma^k Y) \xrightarrow{\times} \Omega^{j+k}(X \times Y)$$

by the formula

(4.2) $[g : M \to U] \times [h : N \to V] = (-1)^{jq}[g \times h : M \times N \to U \times V]$,

where U, V are open neighborhoods of X, Y and g, h are representatives of cobordism classes, as in 2.8; the manifold $M \times N$ has to be taken with the product structure arising from $T(M \times N) = (TM) \times (TN)$. The sign $(-1)^{jq}$ is needed to ensure naturality, commutativity (cf 4.16), and neutral elements. The same formula (4.2) gives rise to relative cross-products

(4.3) $\Omega_\sigma^j(X,A) \times \Omega_\sigma^k(Y,B) \xrightarrow{\times} \Omega_\sigma^{j+k}(X \times Y,\ X \times B \cup A \times Y)$,

for (relatively) closed subsets $A \subset X$, $B \subset Y$.

Interior-(cup-)products

(4.4) $\Omega_\sigma^j(X,A_1) \times \Omega_\sigma^k(X,A_2) \xrightarrow{\smile} \Omega_\sigma^{j+k}(X,A_1 \cup A_2)$ are defined, as usual, by composing 4.3 (where $Y = X$) with the diagonal map $\Delta : X \to X \times X$.

Thus,

(4.5) $a_1 \smile a_2 = \Delta^*(a_1 \times a_2)$, for $a_i \in \Omega_\sigma^*(X,A_i)$.

More geometrically, with representatives $g:M \to U$, $h:N \to U$, we have

(4.6) $[g:M \to U] \smile [h:N \to U] = (-1)^{pj}[M \times_U N \to U]$

provided (g,h) are smooth (where it matters) and tranverse. The σ-structure of $M \times_U N$ is obtained from the trivialization of its normal bundle in $M \times N$ via the (derivative of the) map $\langle g,h\rangle : M \times N \to \mathbb{R}^p$, $\langle g,h\rangle(m,n) = g(m) - h(n)$, as in 2.14 ; this can also be inferred from $[g] \smile [h] = \Delta^*([g] \times [h])$.

The standard properties of $\times$- and $\smile$-products (compare $[D_1]$, VII,7-8) hold and are easily proved. The element $1 = [id:pt \to pt] \in \Omega_\sigma^o(\mathbb{R}^o)$ is neutral with respect to $\times$, the element $1_X = [id : \mathbb{R}^p \to \mathbb{R}^p] \in \Omega_\sigma^o X$ is neutral with respect to $\smile$ in $\Omega_\sigma^* X$. The reader might like some help in proving the commutation law $a_1 \smile a_2 = (-1)^{jk}\, a_2 \smile a_1$; this is provided in (4.16).

Cap-products

(4.7) $(\Omega_\sigma^j X) \times (\Omega^\sigma_{j+k} X) \xrightarrow{\frown} \Omega^\sigma_k X$,

or (if X is compact and $\hat{\Omega}$ denotes Steenrod bordism; 2.31).

(4.8) $(\Omega_\sigma^j X) \times (\hat{\Omega}^\sigma_{j+k} X) \xrightarrow{\frown} \hat{\Omega}^\sigma_k X$

are defined by almost the same formula as $\smile$-products - with some interesting modifications, as follows. Suppose first X is an open subset of $\mathbb{R}^p$. Then (4.7) is indeed defined (as in 4.6) by

$$[g : M \to X] \frown [h : B \to X] = (-1)^{pj}[M \times_X N \to X] ,$$

where (g,h) have to be taken smooth and transverse, and N is compact now, of dimension $j+k$. In this case, there is very little difference then between $\smile$- and $\frown$-products. - A similar description for $\frown$-products works for arbitrary manifolds X if one uses the groups $\Lambda_\sigma^j \cong \Omega_\sigma^j X$, as in 3.3, 3.4.

For general locally compact $X \subset R^p$ the elements $x \in \Omega_\sigma^j X$, $\xi \in \Omega_{j+k}^\sigma X$ have representatives of the form $g : M \to U$, $h : N \to X$, where N is compact, U is an open neighborhood of X in R^p, and g is smooth. We choose a smooth map $h' : N \to U$ transverse to g and sufficiently close to $N \xrightarrow{h} X \hookrightarrow U$. Then $M \times_U N = \{(m,n) \in M \times N \mid g(m) = h'(n)\}$ is a compact σ-structured manifold of dimension k, and $(x \frown \xi) \in \Omega_k^\sigma X$ is represented by $M \times_U N \longrightarrow N \xrightarrow{h} X$; more precisely.

$$(4.9) \qquad [g : M \to U] \frown [h : N \to X] = (-1)^{pj}[M \times_U N \longrightarrow N \xrightarrow{h} X] .$$

For compact $X \subset R^p$ and Steenrod bordism we have to consider a function $d : R^p \to [0,+\infty)$ and $(R \times R^p, (R^p-X)^\natural)$, as in 2.31. For positive $t \in R$, let $(R \times R^p, (R^p-X)^\natural)_{>t}$ denote the part of this pair above t (i.e. first coordinate $> t$); it is homotopy equivalent to $(R \times R^p, (R^p-X)^\natural)$, by linear upwards deformation in R. If C is a closed neighborhood of X in R^p such that $d(z) < \frac{1}{t}$ for $z \in C$ then

$$\Omega_\sigma^*(R \times R^p, (R^p-X)^\natural)_{>t} \cong \Omega_\sigma^*(R \times C, (R^p-X)^\natural \cap R \times C)_{>t} ,$$

by excision 2.12. Cap products (4.8) with Steenrod bordism can now be defined (up to a sign $(-1)^j$) by the following composition

$$(4.10) \qquad \begin{aligned} &(\Omega_\sigma^* C) \times (\hat{\Omega}_*^\sigma X) \cong \Omega_\sigma^*(R\times C)_{>t} \times \Omega_\sigma^*(R\times C, (R^p-X)^\natural \cap R\times C)_{>t} \\ &\xrightarrow{\smile} \Omega_\sigma^*(R\times C, (R^p-X)^\natural \cap R\times C)_{>t} \cong \hat{\Omega}_*^\sigma X , \end{aligned}$$

and passing to $\varinjlim$ over the directed set of closed neighborhoods C. This looks complicated but it acquires a simple geometric meaning if we use our geometric description of Steenrod bordism (cf.2.31):

An element $\xi \in \hat{\Omega}^{\sigma}_{j+k}X$ is then represented by a variable smooth (j+k)-dimensional compact $h_t : N_t \to R^p$, closing in on X as $t \to +\infty$; it suffices to know N_t for sufficiently large t (this is the significance of the restriction to $>t$ above). An element $x \in \Omega^j_{\sigma}X$ is represented by a proper $g : M \to U$ over some neighborhood U of X ; for sufficiently large t (say $t > \tau$) , $h_t(N_t) \subset U$. We can assume that $\{h_t\}_{t>\tau}$ and $id \times g :: R_{>\tau} \times M \to R_{>\tau} \times U$ are transverse. Then $\pm x \frown \xi \in \hat{\Omega}^{\sigma}_k X$ is represented by the variable smooth manifold $M \times_U N_t \longrightarrow U \subset R^p$. - This is very similar again to (4.6). With the necessary precautions, as indicated by the preceding explanations we can write

(4.11) $[g : M \to U] \frown [\{h_t : N_t \to U\}_{t>\tau}] = (-1)^{pj}[\{M \times_U N_t \longrightarrow U\}_{t>\tau}]$,

where $j = \dim M - p$.

(4.12) Comment. Proper σ-structured maps $\alpha : Y \to X$ between manifolds play a double role in our treatment of cobordism: On the one hand they are treated as morphisms in the underlying category, inducing homomorphisms on Ω_{σ} , on the other hand they are treated as (representatives of) cobordism elements $[\alpha] \in \Omega_{\sigma}$. Exploiting this double role gives easy (almost tautological) proofs of some formulas in cobordisms. For some examples, consider pullback diagrams as in (2.19), i.e.

$$
\begin{array}{ccccc}
L & \xrightarrow{\beta'} & N & \xrightarrow{\alpha'} & M \\
\downarrow{g''} & & \downarrow{g'} & & \downarrow{g} \\
Z & \xrightarrow[\beta]{} & Y & \xrightarrow[\alpha]{} & X
\end{array} \tag{4.13}
$$

where now α, β, g are proper and σ-structured smooth maps, (α, g) and (β, g') are transverse, both squares (and hence the composite) are pullback squares with the induced σ-structures on α', β', g', g''. Ignoring signs (which are treated in 4.16 and in the proof of 2.18, with more care) we have $[g'] = \alpha^*[g]$ and $\alpha_![g'] = [\alpha g'] =$ $[\alpha] \smile [g]$, hence

$$(4.14) \qquad \alpha_!\alpha^*[g] = [\alpha] \smile [g] \text{ , or } \alpha_!\alpha^* = [\alpha]\smile .$$

More generally, $\alpha_!([\beta] \smile \alpha^*[g]) = \alpha_!([\beta] \smile [g']) = \alpha_!([\beta g"]) =$ $=[\alpha\beta g"] = [\alpha\beta] \smile [g] = \alpha_![\beta] \smile [g]$.

$$(4.15) \qquad \alpha_!(y \smile \alpha^* x) = \pm\, \alpha_!(y) \smile x \text{ , for } x \in \Omega_\sigma^* Y \text{ , } y \in \Omega_\sigma^* Y .$$

For $y = 1_Y$ this reduces to 4.14.

In the same spirit, $g^*\alpha_![\beta] = g^*[\alpha\beta] = [\alpha'\beta'] = \alpha'_![\beta'] =$ $= \alpha'_! g'^*[\beta]$. This proves 3.10; it did not use properness of g .

(4.16) <u>Commutativity of $\times$- and $\smile$-products</u>. It is enough to consider $(\Omega_\sigma^j X)\times(\Omega_\sigma^k Y) \xrightarrow{\times} \Omega_\sigma^{j+k}(X \times Y)$ for open sets $X \subset R^p$, $Y \subset R^q$. Other spaces X,Y can then be handled by passing to $\varinjlim$; or by looking at representatives over euclidean open neighborhoods. Commutativity of $\smile$-products is obtained by composing with the diagonal. For relative groups $\Omega_\sigma^*(X,A)$ etc. the proofs are similar to the ones for absolute groups.

Let $[p : M \to U] \in \Omega_\sigma^j X$, $[q : N \to V] \in \Omega_\sigma^k Y$ as in 4.2, and μ,ν the given σ-structures of M,N . Then the switch-map $\tau' : N\times M \to M\times N$ (which is covered by the bundle map $TN\times TM \to TM\times TN$) takes $\mu\times\nu$ into $(-1)^{|M||N|}\nu\times\mu$, where $|M| = p-j$, $|N| = q-k$. The diagram

$$(4.17) \qquad \begin{array}{ccc} N \times M & \xrightarrow{\tau'} & M \times N \\ \downarrow{\scriptstyle h\times g} & & \downarrow{\scriptstyle g\times h} \\ V \times U & \xrightarrow{\tau} & U \times V \end{array}$$

is a smooth pullback diagram (we assume g,h smooth). Therefore

$\tau^*[g\times h]$ is (cf. 2.16) represented by $h\times g : N\times M \longrightarrow V\times U$; we have only to determine the σ-structure of $N \times M$. In order to do so we have to represent $R^p \times R^q \times N \times M$ as a tubular neighborhood of the pullback in $V \times U \times M \times N \subset R^q \times R^p \times M \times N$. Such a representation ρ is given by $\rho : (x,y,n,m) \longmapsto (h(n)+y,g(m)+x,m,n)$. This map ρ takes the σ-structure $s^q \times s^p \times \mu \times \nu$ of $R^q \times R^p \times M \times N$ into $(-1)^{pq+|M||N|} s^p \times s^q \times \nu \times \mu$. It follows (2.16) that $\tau^*[g\times h] = (-1)^{pq+|M||N|} [h\times g]$, where $M \times N$, $N \times M$ are structured by $\mu \times \nu$, $\nu \times \mu$. From 4.2 it now follows that

$$\tau^*([g] \times [h]) = (-1)^{pq+|M||N|+jq}[h\times g]$$

$$= (-1)^{pq+|M||N|+jq+kp}[h] \times [g] = (-1)^{jk}[h] \times [g] ,$$

as required. ∎

5. The fixed point transfer for cobordism groups Ω_σ^*.

(5.1) This notion can be thought of as fixed point invariant which is attached to a continuous family of continuous maps. More precisely, it is defined in the following situation. Let $p : E \to B$ denote a map which is ENR_B (= euclidean neighborhood retract over B ; cf [D_4]) ; for instance, p could be a fibre bundle with polyhedral fibre. Let $W \subset E$ an open subset and $f : W \to E$ a fibre-preserving map ($pf = p|W$) such that $Fix(f) = \{w \in W \mid fw = w\}$ is proper over B ($p|Fix(f)$ is proper); we say that f is <u>compactly fixed</u>. For every cohomology theory h over B the fixed point transfer t_f is then a natural homomorphism

$$t_f : \check{h}(Fix(h)) \longrightarrow hB$$

with interesting properties and applications (cf. [B-G], [D_4]). We shall now give a simple geometric interpretation of $t_f : \Omega_\sigma^*(Fix(f)) \longrightarrow \Omega_\sigma^* B$ - at least if B is a locally compact subset of some R^p(the general theory only assumes B metric). Roughly speaking, t_f turns out to be a Gysin-homomorphism on suitable smooth σ-structured approximations of $Fix(f)$.

<u>(5.2) Reduction to $E = R^n \times B$</u> . By definition of an ENR_B there is an open set $\mathcal{O}$ in some $R^n \times B$ and maps <u>over B</u> (i.e. fibre-preserving)

$$i : E \longrightarrow \mathcal{O} \subset R^n \times B \ , \quad r : \mathcal{O} \longrightarrow E \ .$$

such that $ri = id$. The composite map

$$r^{-1}W \xrightarrow{r} W \xrightarrow{f} E \xrightarrow{i} R^n \times B$$

then has the same fixed point set, $Fix(ifr) = Fix(f)$, and (cf. [D_4]) the same transfer as f . We can therefore replace f by ifr , i.e. we can (without loss of generality) assume that

$E = R^n \times B$, and W an open subset of $R^n \times B$; thus

$$(5.3) \quad \begin{array}{l} f : W \longrightarrow R^n \times B \quad , \quad f(z,b) = (\varphi(z,b),\, b) \; , \\ \text{where} \quad \varphi : W \longrightarrow R^n \; . \end{array}$$

The fixed point transfer t_f depends only on the germ of f around $\mathrm{Fix}(f)$; it is unchanged if we restrict f to any neighborhood of $\mathrm{Fix}(f)$ in W . By shrinking W (if necessary) we can therefore assume that f resp. φ is defined on the closure $\overline{W}$ (but $\mathrm{Fix}(f) \subset W$) . Furthermore, B being locally compact in R^p it has an open neighborhood V in R^p in which it is closed. Therefore $\overline{W}$ is a closed subset of $R^n \times V$. By Tietze's extension lemma the map $\varphi : \overline{W} \to R^n$ (and hence f) has a continuous extension φ' to all of $R^n \times V$; thus

$$f' : R^n \times V \longrightarrow R^n \times V \; , \; f'(z,v) = (\varphi'(z,v),v)$$
$$f' \mid \overline{W} = f \; .$$

Of course, $\mathrm{Fix}(f') \longrightarrow V$ need not be proper, but (as in $[D_4]$, 8.6) we have the following

<u>(5.4) Lemma</u>. <u>There are open neighborhoods V' of B in V and W' of $\mathrm{Fix}(f)$ in $R^n \times V'$ such that $f'|W' : W' \longrightarrow R^n \times V'$ is compactly fixed, i.e. $\mathrm{Fix}(f'|W') \longrightarrow V'$ is proper</u> (and, of course, $f'(z,b) = f(z,b)$ if $(z,b) \in (W \cap W')$.

<u>Proof</u>. Since $\mathrm{Fix}(f) \longrightarrow B$ is proper there is ($[D_4]$,1.3) a continuous function $\rho : B \longrightarrow (0,+\infty)$ such that $(z,b) \in \mathrm{Fix}(f) \Rightarrow \|z\| < \rho(b)$. By Tietze's lemma, we can extend ρ to V , i.e. we can assume $\rho : V \longrightarrow (0,+\infty)$. Let Y an open neighborhood of $\mathrm{Fix}(f)$ in $R^n \times V$ such that $\overline{Y} \cap (R^n \times B) \subset W$; we can choose Y such that $\overline{Y} \to V$ is proper - because if this is not already the case we intersect Y with the tube

$\{(z,v) \in \mathbb{R}^n \times V \mid \|z\| < \rho(v)\}$. Let K be a closed neighborhood of $\mathrm{Fix}(f)$ in $\mathbb{R}^n \times V$ such that $K \subset Y$; thus
$\mathrm{Fix}(f) \subset \mathring{K} \subset K \subset Y \subset \overline{Y} \subset \mathbb{R}^n \times V$.

The set $\overline{Y} - \mathring{K}$ is proper over V and contains no fixed point of f ; thus $z \neq \varphi(z,v)$ if $v \in B$ and $(z,v) \in \overline{Y} - \mathring{K}$. Therefore the set

$$V' = \{v \in V \mid z \neq \varphi(z,v) \quad \text{for all} \quad (z,v) \in \overline{Y} - \mathring{K}\}$$

is open in V and contains B . Let $W' = Y \cap (\mathbb{R}^n \times V')$. Then $\mathrm{Fix}(f'|W')$ is contained in K and is closed in W' hence closed in $K \cap (\mathbb{R}^n \times V')$. But $K \cap (\mathbb{R}^n \times V') \longrightarrow V'$ is proper (because $K \to V$ is) , hence $\mathrm{Fix}(f'|W') \longrightarrow V'$ is proper. ∎

(5.5) Reduction to the case where B is open in $\mathbb{R}^p$. We assume a compactly fixed map $f : W \longrightarrow \mathbb{R}^n \times B$, where W is open in $\mathbb{R}^n \times B$, and B is locally compact in $\mathbb{R}^p$; the general case was reduced to this situation (in 5.2). We have (cf. 5.3,5.4) extended f to a compactly fixed $f' : W' \longrightarrow \mathbb{R}^n \times V$, where V is an open neighborhood of B in $\mathbb{R}^p$, and W' is an open neighborhood of $\mathrm{Fix}(f)$ in $\mathbb{R}^n \times V$. Let U be any open neighborhood of $\mathrm{Fix}(f)$ in W' . Then

(5.6) $$V_U = \{v \in V \mid \mathrm{Fix}(f')_v \subset U\}$$

(where $\mathrm{Fix}(f')_v = \mathrm{Fix}(f') \cap (\mathbb{R}^n \times \{v\})$) is an open neighborhood of B in V (because $\mathrm{Fix}(f') \to V$ is proper, hence closed).
The set $U' = U \cap (\mathbb{R}^n \times V_U)$ is also an open neighborhood of $\mathrm{Fix}(f)$ in W' , and

(5.7) $$f'_U = f'|U' \ : \ U' \longrightarrow \mathbb{R}^n \times V_U$$

is a compactly fixed extension of f . By definition $([D_4],3.2)$ and naturality $([D_4],3.12)$ of the transfer we can obtain

$t_f : \check{h}\,\mathrm{Fix}(f) \longrightarrow hB$ by passing to $\varinjlim$ (over $\{U\}$) with the direct system of homomorphisms

(5.8) $$hU' \xrightarrow{j^*} \check{h}\,\mathrm{Fix}(f'_U) \xrightarrow{t_{f'_U}} h(V_U) \xrightarrow{i^*} hB ,$$

where i,j are inclusions. Explicitely, <u>every $x \in \check{h}\,\mathrm{Fix}(f)$ has a representative of the form $u \in hU'$, and $t_f(x) = i^* t_{f'_U} j^* u$.</u> In order to know or to describe t_f it suffices therefore to know or describe $t_{f'_U}$, or even $(t_{f'_U} j^*)$. The progress with $t_{f'_U}$ is, of course, that f'_U is over an <u>open</u> set (V_U) of R^p; the progress with $t_{f'_U} j^*$ is that we can look at the (good) open set U' in $R^n \times R^p$ instead of the (possibly bad) set $\mathrm{Fix}(f'_U)$.

<u>(5.9) The transfer in case $B \subset R^p$ is open, $E = R^n \times B$.</u> We have open subsets $W \subset (R^n \times B) \subset (R^n \times R^p)$ and maps

$$f : W \longrightarrow R^n \times B , \quad \varphi : W \to R^n , \quad f(z,b) = (\varphi(z,b),b) ,$$

such that $\mathrm{Fix}(f) \longrightarrow B$ is proper. Moreover, we have an open neighborhood U of $\mathrm{Fix}(f)$ in $R^n \times B$, and we have to describe the composite

$$t_f^U : hU \xrightarrow{j^*} \check{h}\,\mathrm{Fix}(f) \xrightarrow{t_f} hB ,$$

in particular if $h = \check{h} = \Omega_\sigma^*$. We can assume $U = W$ - otherwise we just replace f by $f \mid W \cap U$. Let

$$\varphi_* : W \longrightarrow R^n , \quad \varphi_*(z,b) = z - \varphi(z,b) ,$$

hence $\mathrm{Fix}(f) = \varphi_*^{-1}(0)$. We can approximate φ_* by a smooth map $\varphi_o : W \to R^n$ for which $0 \in R^n$ is a regular value. Then $\varphi_o^{-1}(0)$ is a smooth p-dimensional submanifold of $W \subset R^n \times R^p$ with trivialized (by the derivative of φ_o) tubular neighborhood; this makes it canonically σ-structured. For this, φ_o need only be smooth in a neighborhood of $\varphi_o^{-1}(0)$; we can choose it such that $\varphi_o|W\text{-}N = \varphi_*|W\text{-}N$, where N is any prescribed neighborhood of $\mathrm{Fix}(f)$ in W. If we choose N small enough then $\varphi_o^{-1}(0) \to B$ is still proper (just make $\bar{N}$ proper), and even the deformation $\varphi_o \simeq \varphi_*$ has proper counterimage of 0. It follows that $\varphi_o^{-1}(0) \to B$ represents a well defined (independent of the choice of φ_o) cobordism class; this is the <u>index of f</u>,

$$(5.10) \qquad \mathrm{Index}(f) = I(f) = [\varphi_o^{-1}(0) \to B] \in \Omega_\sigma^o B \quad {}^{*)} .$$

Roughly speaking, <u>the index of f is the fixed point set $\mathrm{Fix}(f)$</u> viewed (after approximation) as a proper p-dimensional manifold over B, with σ-structure induced by (id-f). The σ-structure can be be thought of as a "multiplicity" with which the manifold $\varphi_o^{-1}(0)$ is to be taken. - If σ stands for "stable trivialization of the normal bundle" then Ω_σ^* is stable cohomotopy, hence $I(f) \in \pi^o_{\mathrm{stable}}(B \oplus pt)$.

<u>The transfer t_f, or rather the composite</u>

$\Omega_\sigma^* W \xrightarrow{j^*} \Omega_\sigma^* \mathrm{Fix}(f) \xrightarrow{t_f} \Omega_\sigma^* B$ <u>coincides with the composite</u>

$$(5.11) \qquad \Omega_\sigma^* W \xrightarrow{j^*} \Omega_\sigma^*(\varphi_o^{-1}(0)) \xrightarrow{(p|\varphi_o^{-1}(0))_!} \Omega_\sigma^* B .$$

Explicitely, an element $x \in \Omega_\sigma W$ is represented by a proper σ-structured manifold $g : M \to W$; this can be taken smooth and transverse to $i : \varphi_o^{-1}(0) \longrightarrow W$. Then

$$(5.12) \qquad t_f\, j^*(x) = \pm [g^{-1}\varphi_o^{-1}(0) \xrightarrow{pg} B] ;$$

---

*) Such an interpretation of $I(f)$ was suggested by the author in Math. Zeitschr. 25 (1974) on p. 297, section (5.5). Details have been carried out by T. Koźniewski, Warsaw. (Oral communication April 1976, letter July 1977.)

the sign has to be chosen as explained in 2.14. In fact, the same description applies to $t_f(y)$ for $y \in \Omega_\sigma^* \operatorname{Fix}(f)$ because every such y has a representative $g : M \to W$ for a suitable neighborhood W of $\operatorname{Fix}(f)$; with φ_o chosen for this $W = W(y)$ one has

$$(5.12') \qquad t_f(y) = \pm \, [g^{-1}\varphi_o^{-1}(0) \xrightarrow{pg} B] \,.$$

Roughly speaking, <u>the transfer is the Gysin homomorphism induced by $\operatorname{Fix}(f) \longrightarrow B$</u> when $\operatorname{Fix}(f)$ is viewed (after approximation) as a σ-structured manifold.

This description of $t_f j^*$ resp t_f also makes sense and is correct for other cohomology theories h (which are not necessarily of the form Ω_σ^*) . Indeed, the trivialization of the normal bundle gives an isomorphism (suspension isomorphism)

$$h^j(\varphi_o^{-1}(0)) \;\cong\; h^{j+n}(\mathbb{R}^n \times B \,,\, \mathbb{R}^n \times B - \varphi_p^{-1}(0)) \,.$$

Since $\varphi_o^{-1}(0)$ is proper over B it is contained in a tube N_ρ of some radius $\rho = \rho(b)$, $N_\rho = \{(z,b) \mid \|z\| < \rho(b)\}$; this gives rise to another suspension isomorphism $h^j B \cong h^{j+n}(\mathbb{R}^n \times B \,,\, \mathbb{R}^n \times B - N_\rho)$. The Gysin-homomorphism for h is then the composite

$$h^j\varphi_o^{-1}(0) \cong h^{j+n}(\mathbb{R}^n \times B, \mathbb{R}^n \times B - \varphi_o^{-1}(0)) \longrightarrow h^{j+n}(\mathbb{R}^n \times B, \mathbb{R}^n \times B - N_\rho) \cong h^j B \;,$$

and with this Gysin homomorphism the description of $t_f j^*$ is as in 5.11. The transfer t_f itself is obtained by passing to $\varinjlim$ with respect to $\{W\}$.

<u>(5.13) Examples</u>. (i) Let $B = \mathbb{C} - \{0\}$, $E = \mathbb{C} \times B$, and $f : E \to E$, $f(z,b) = (0,b)$, hence $\varphi : E \to \mathbb{C} = \mathbb{R}^2$, $\varphi(z,b) = 0$, $\varphi_o(z,b) = z$. We have $\varphi_o^{-1}(0) = \{0\} \times B \approx B$ with the obvious trivialization of the normal bundle. Thus $\varphi_o^{-1}(0) \to B$ is the identity map of B , $I(f) = 1$, $t_f = \mathrm{id}$.

(ii) Let B, E as above but $f(z,b) = (z-bz,b)$, hence $\varphi_o(z,b) = bz$.

As before, $\varphi_o^{-1}(0) = \{0\} \times B$ but the normal bundle of $\varphi_o^{-1}(0)$ is trivialized differently: the trivializing 2-frame makes one full (2π-) turn as $b \in \mathbb{C}-\{0\}$ moves around the unit circle once. In ordinary cohomology or oriented cobordism (example 1.4 ii) one still has $I(f) = 1$, $t_f = id$, but in stably parallelized cobordism (example 1.4 iv), where $\Omega_\sigma^o(\mathbb{C}-0) = \Omega_\sigma^o(S^1) = \pi_{st}^o(S^1 \oplus pt) = \mathbb{Z} \oplus {}^{\mathbb{Z}}/2\mathbb{Z}$ one finds that $I(f) \in \Omega_\sigma^o(\mathbb{C}-0) = \mathbb{Z} \oplus {}^{\mathbb{Z}}/2\mathbb{Z}$ is the element $(1,\overline{1})$. Thus $\tilde{I} = (I(f)-1) \neq 0 \smile$, although $2\tilde{I} = 0$ and $\tilde{I} \smile \tilde{I} = 0$. The transfer t_f equals $I(f)\smile$, in this case.

(iii) Let $B = \mathbb{C}-\{0\}$ as above but $E = (\mathbb{C}-\{0\}) \times B$, and $f(z,b) = \left(\frac{bz}{\|bz\|}, b\right)$, hence $\varphi_o(z,b) = z - \frac{bz}{\|bz\|}$. The manifold $\varphi_o^{-1}(0) = \{(z,b) \mid \|z\| = 1 \text{ and } b > 0\} \approx S^1 \times (0,+\infty)$ is still diffeomorphic to $\mathbb{C} - \{0\}$ and it has the same (twisted) trivialization of the normal bundle as in (ii), but the projection $\varphi_o^{-1}(0) \to B$ is not the same as in (ii). In fact, one finds $I(f) = 0$ in oriented cobordism, and $I(f) = (0,\overline{1}) \neq 0$ in stably parallelized cobordism.

<u>(5.14) Example</u>: <u>The index $I(G)$ of a compact Lie group.</u> Let $B = G$ a compact (connected) Lie group of dimension p, $E = G \times B$, and $f : E \to E$, $f(g,b) = (b\cdot g, b)$. Clearly $\mathrm{Fix}(f) = G \times \{e\} \approx G$, where $e \in G = B$ is the neutral element. The index $I(f) \in \Omega_\sigma^o G = \pi_{stable}^o(G \oplus pt)$ is represented by the constant map $G \to \{e\} \subset B$, where G has to be taken with a suitable σ-structure, i.e. with a trivialization of the stable normal bundle $T'G$. If ⓔ is a (small) euclidean neighborhood of e we can think of $I(f)$ as an element of $\Omega_\sigma^o(G, G-ⓔ)$, or even $\Omega_\sigma^o(G,G-\{e\}) \cong \Omega_\sigma^{-p}(pt) = \Omega_p^\sigma(pt) = \pi_p^{stable}$ = p-th stable homotopy (of the 0-sphere). As such, $I(f)$ is denoted by $I(G) \in \pi_o^{stable}$. By excision $\Omega_\sigma^o(G,G-\{e\} \cong \Omega_\sigma^o(U,U-\{e\})$, where $U \approx \mathbb{R}^p$ is a coordinate neighborhood of e in G. We can

therefore replace B by $U \approx R^p$ for studying $I(G)$.

As in (5.2), we embed $G \subset R^n$ where it is a retract of its tubular neighborhood, say $\pi : \nu G \to G$, and we extend f to $(\nu G) \times B$ resp. $(\nu G) \times U$ by $f(x,y) = (y \cdot \pi(x), y)$; the fixed point set is still $G \times \{e\}$. (We should also embed B into euclidean space, as in (5.5), but since we restrict attention to $U \approx R^p$ this is redundant.) We now determine the trivialization of the normal bundle T' of $G \times \{e\}$ in $R^n \times U = R^n \times R^p = R^{n+p}$. We have

(5.15) $\quad \varphi_* = \varphi_o : (\nu G) \times U \longrightarrow R^n$, $\varphi_o(x,y) = x - y \cdot \pi(x)$. It is easy to see (compare also 5.16) that 0 is a regular value, so that T' is trivialized by the derivative $\dot{\varphi}_o$ of φ_o at $G \times \{e\} = \varphi_o^{-1}(0)$. In order to calculate $\dot{\varphi}_o$ at a point $(g,e) \in G \times \{e\}$ we write $y \cdot \pi(x) = (y \cdot \pi(x) \cdot g^{-1} \cdot g = R_g(y \cdot R_{g^{-1}}(\pi(x)))$, where R_g is right translation with g. We apply the chain rule using the (obvious) fact that the derivative of the multiplication $U \times U \to G$ at (e,e) is $(\eta_1, \eta_2) \longmapsto \eta_1 + \eta_2$, and get

(5.16) $\quad \dot{\varphi}_o(\xi,\eta) = \xi - \dot{R}_g(\eta + \dot{R}_{g-1}\,\dot{\pi}(\xi)) = (\xi - \dot{\pi}(\xi)) - \dot{R}_g(\eta)$,

for tangent vectors $\xi \in T_g(\nu G)$, $\eta \in T_e U = R^p$; notice that $\xi - \dot{\pi}(\xi)$ is the component of ξ along the fibre $\pi^{-1}(g)$.

If we replace f by $f'(g,b) = (g \cdot b^{-1}, b)$ then $\mathrm{Fix}(f') = \mathrm{Fix}(f) = G \times \{e\}$, of course; the formula (5.16) which determines the trivialization of the normal bundle becomes

(5.16') $\quad \dot{\varphi}_o'(\xi,\eta) = (\xi - \dot{\pi}(\xi)) + \dot{L}_g(\eta)$,

where L_g = left translation with g. But f and f' have the same index, $I(f') = I(f)$, because the homeomorphism $(g,b) \longmapsto (g^{-1}, b)$ takes one into the other (cf. [D_4], 3.12) so that the two trivializations 5.16, 5.16' of the normal bundle of

$G \times \{e\}$ lead to the same element $I \in \Omega_\sigma^o(G,G-\{e\})$.

This index $I(G) \in \pi_p^{stable} \cong \Omega_\sigma^o(G,G-\{e\})$ can be described differently and has been studied by Atiyah-Smith [A-S] , Becker-Schultz [B-S] , and others. That the two descriptions really concern the same element of π_p^{stable} was pointed out to me by Th. Bröcker, in a letter; the formula (5.16) which easily gives the result (cf. below) is essentially a translation of the proof of ([B-S], 3.5). The other description is in terms of the Pontrjagin-Thom element $[Y,\rho] \in \Omega_\sigma^{-p}(pt)$ of a parallelized compact p-dimensional smooth manifold Y , where $\rho = (\rho_y : R^p \cong T_yY)_{y \in Y}$ denotes the parallelization. It is obtained as follows: Embed $Y \subset R^n$ with (open) tubular neighborhood (= normal bundle) $\pi : \nu Y \to Y$. Then $(\nu Y) \times R^p$ is an open subset of $R^n \times R^p$, and the composite map

$$\psi : (\nu Y) \times R^p \xrightarrow{1 \times_Y \rho} (\nu Y) \oplus TY \cong R^n \times Y \xrightarrow{proj} R^n$$

has regular value 0 , $\psi^{-1}(0) = Y \times \{0\}$, and the derivative $\dot\psi$ at $\psi^{-1}(0)$ is given by $\dot\psi(\xi,\eta) = (\xi - \dot\pi(\xi)) + \rho_y(\eta)$. Therefore $Y = Y \times \{0\}$, with this trivialization of its normal bundle in $R^n \times R^p$, defines an element in $\Omega_\sigma^{-p}(pt)$, denoted by $[Y,\rho]$. - In particular, if $Y = G$ is a compact Lie group with parallelization $\rho_y = \dot L_y : R^p = T_eG \xrightarrow{\cong} T_yG$ then the formula for $\dot\psi$ clearly coincides with 5.16', hence $[Y,\rho] = I(G)$. ∎

References

[A-R] R. Abraham - J. Robbin, Transversal Mappings and Flows. Benjamin, New York, 1967

[A] M.F. Atiyah, K-Theory. Benjamin, New York, 1967

[A-S] M.F. Atiyah - L. Smith, Compact Lie Groups and the Stable Homotopy of Spheres. Topology 13 (1974) 135-142

[B-G] J.C. Becker - D.H. Gottlieb, Transfer Maps for Fibrations and Duality. Comp. Math. 33 (1976) 107-133

[B-S] J.C. Becker - R. E. Schultz, Fixed Point Indices and Left Invariant Framings. To appear, Proc. Conf. on Homotopy Theory, Evanston 1977, in Springer Lecture Notes Math.

[B] E.H. Brown, Abstract Homotopy Theory. Trans.Am.Math.Soc 119 (1965) 79-85

[B-R-S] S. Buoncristiano - C.P. Rourke - B.J. Sanderson, A Geometric Approach to Homology Theory. Cambridge Univ. Press, London 1976

[D_1] A. Dold, Lectures on Algebraic Topology. Springer, Heidelberg 1972

[D_2] ----------- The K-Theory and Cobordism Theory. Associated with a General Cohomological Structure. Conf. on Topol. and its Applications Budva (Yugoslavia) 1972

[D_3] ----------- Chern Classes in General Cohomology. Symp.Math.INDAM V (1970) 385-410

[D_4] ----------- The Fixed Point Transfer of Fibre-Preserving Maps. Math.Z.148 (1976) 215-244

[H] M.W. Hirsch, Differential Topology.
Grad. Texts in Math.; Springer,
Heidelberg, 1976

[K-K-S] D.S. Kahn - J. Kaminker - C. Schochet, Generalized Homology
Theories on Compact Metric Spaces.
Mich.Math.J. 24 (1977) 203-224

[Q] D. Quillen, Elementary Proofs of Some Results of Cobordism
Theory using Steenrod Operations.
Advances in Math. 7 (1971) 29-56

[S] R.E. Stong, Notes on Cobordism Theory.
Math. Notes, Princeton Univ. Press 1968

[W] H. Whitney, Geometric Methods in Cohomology Theory.
Proc. Nat. Acad. Sci. USA, 33(1947) 7-9.

Added in proof:

A. Bojanowska - S. Jackowski, Geometric Bordism and Cobordism.
Lect. Notes VIth Alg. Topol. Summer School, Gdansk 1973
Polish Acad. Sci.

(these notes also contain an exposition of geometric cobordism, somewhat closer to Quillen's paper [Q] and less elementary than ours).

IMMERSIONS IN MANIFOLDS OF POSITIVE WEIGHTS

Henry Glover, Bill Homer and Guido Mislin

Introduction

In this note we generalize the main result of [GH] in two ways. First, we extend the result to cover certain manifolds which are not nilpotent, and obtain immersion results for arbitrary (generalized) spherical space forms. Second, we show that for immersions into manifolds of positive weights, the theorem 1.1 of [GH] takes a particularly simple form to the extent that condition (iv) stated there, which is hard to verify, becomes redundant. We give some applications to Grassmann manifolds.

Our results follow: Let M and N^n be compact smooth $\mathbb{Z}/2\mathbb{Z}$ - good manifolds (in the sense of [BK]) and let V be a $\mathbb{Z}/2\mathbb{Z}$ - good smooth manifold. Let W^{n+k} be a nilpotent smooth manifold. (Note that M, V need not have the same dimensions as N, W respectively.) Denote by X_p, respectively $\hat{X}_p$, the Bousfield-Kan localization, respectively completion of X at the prime p (cf. [BK]). Let $\hat{X}_{odd} = \times\{\hat{X}_p : p > 2 \text{ is prime}\}$.

Theorem 0.1 Suppose

(i) there exists an immersion $j : M \to V$

(ii) there exist homotopy equivalences

$$\lambda: \hat{N}_2 \to \hat{M}_2 \quad \text{and} \quad \mu: \hat{V}_2 \to \hat{W}_2 ,$$

(iii) $k \geq [n/2] + 1$ is odd, and

(iv) there exist maps $\hat{i}_{odd} : N \to \hat{W}_{odd}$ and $i_0 : N \to W_0$ such that the diagram

$$\begin{array}{ccc} N & \xrightarrow{\{\hat{i}_{odd}, \hat{i}_2\}} & \hat{W}_{odd} \times \hat{W}_2 = \hat{W} \\ {\scriptstyle i_0}\downarrow & & \downarrow{\scriptstyle can} \\ W_0 & \xrightarrow{can} & (\hat{W}_{odd} \times \hat{W}_2)_0 = (\hat{W})_0 \end{array}$$

commutes (up to homotopy), where $\hat{i}_2$ is the composition

$$N \xrightarrow{can} \hat{N}_2 \xrightarrow{\lambda} \hat{M}_2 \xrightarrow{\hat{j}_2} \hat{V}_2 \xrightarrow{\mu} \hat{W}_2 .$$

Then N^n immerses in W^{n+k} (denoted $N \subseteq W$).

In many applications the homotopy set $[N, (\hat{W})_0]$ consists of the constant map only, so that (iv) is immediate. This is for instance the case in the following corollary.

<u>Corollary 0.2</u> Let G be a finite group which acts freely and smoothly on a homotopy sphere Σ^{2n+1} and suppose that G possesses a nilpotent subgroup H of odd index whose action extends to an action on a pair $\Sigma^{2n+1} \subset \Sigma^{2n+2s+1}$ for some $s \geq 0$. Then

(i) $\Sigma^{2n+1}/G \subseteq (\Sigma^{2n+2s+1}/H) \times \mathbb{R}^{2t+1}$ if $t \geq 0$ is such that $2s + 2t \geq n$.

(ii) $(\Sigma^{2n+1}/G) \times \mathbb{R} \subseteq (\Sigma^{2n+2s+1}/H) \times \mathbb{R}^{2t}$ if $t \geq 0$ is such that $2s + 2t \geq n + 3$.

In particular we obtain

<u>Corollary 0.3</u> Every (generalized) smooth spherical space form Σ^{2n+1}/G with G of odd order immerses in metastable euclidean space $\mathbb{R}^{2[3(n+1)/2]}$.

The corollary applies for instance to the (non-nilpotent) manifolds Σ^{2n+1}/G considered by Petrie [P], with G metacyclic of odd order. It was shown in [GH] that Corollaries 0.2 and 0.3 are nearly best possible. We also could have obtained Corollary 0.3 as a special case of the following.

<u>Corollary 0.4</u> Suppose that M^n is a π-manifold and assume $\hat{M}_2 \simeq \hat{N}_2$. Then $N_n \subseteq \mathbb{R}^{n+2[(n+2)/4]+1}$.

The following is a variation of Theorem 0.1, adapted to immersions in manifolds of <u>positive weights</u> (cf. section 1).

<u>Theorem 0.5</u> Suppose

(i) there exists an immersion $j : M \to V$

(ii) there exist homotopy equivalences

$\zeta : N_2 \to M_2$ and $\sigma : V_2 \to W_2$

(iii) $k \geq [n/2] + 1$ is odd, and

(iv) W has positive weights.

Then $N^n \subseteq W^{n+k}$.

We will apply this theorem to Grassmann manifolds. Let $\mathbb{R}G_{u,v}$ denote the manifold of u-dimensional linear subspaces of $\mathbb{R}^{u+v}$. Suppose uv is odd. Idnetifying $\mathbb{R}^{u+v}$ with $C^{(u+v)/2}$, we can define a linear action of C_n, a cyclic group of order n, on $\mathbb{R}G_{u,v}$ by acting with a primitive n-th root of 1 on the coordinates of $C^{(u+v)/2}$. The action of C_n on $\mathbb{R}G_{u,v}$ is free if n is odd, since then an invariant linear subspace of $\mathbb{R}^{u+v}$ has necessarily an even dimension.

<u>Corollary 0.6</u> Suppose $u,v,r,s,w \geq 0$ are integers such that u and v are odd and r and s are even. Let $m = (u+r)(v+s)+w$.

(i) If w is odd and $m \geq [3uv/2] + 1$, then

$$\mathbb{R}G_{u,v}/C_{2t+1} \quad \mathbb{R}G_{u+r,v+s} \times \mathbb{R}^w$$

(ii) If w is even and $m \geq [3(uv+1)/2] + 1$, then

$$\mathbb{R}G_{u,v}/C_{2t+1} \times \mathbb{R} \quad \mathbb{R}G_{u+r,v+s} \times \mathbb{R}^w.$$

1. Spaces with positive weights

A finite nilpotent CW-complex X is said to have <u>positive weights</u>, if there exists a rational equivalence $\phi: X \to X$ and an integer $d > 1$ such that

$$\phi_* = 0 : \tilde{H}_*(X;\mathbb{Z}/d) \to \tilde{H}_*(X;\mathbb{Z}/d)$$

This condition is equivalent to various other conditions on X, and in particular to the following one which we use for Lemma 1.4. For each pair of distinct primes p, q there is a p-equivalence $\theta: X \to X$ inducing 0 in $\tilde{H}_*(X;\mathbb{Z}/q)$ (cf. [BS]).

<u>Lemma 1.1</u> If $\mathbb{R}G_{u,v}$ is nilpotent, then $\mathbb{R}G_{u,v}$ has positive weights.

<u>Proof</u> For $\mathbb{R}G_{u,v}$ nilpotent and q any sufficiently large prime, it was shown in [F] that there exists a self map $\phi^q: \mathbb{R}G_{u,v} \to \mathbb{R}G_{u,v}$ such that for every i, $H^i(\mathbb{R}G_{u,v};Q)$ has a basis of eigenvectors of $(\phi^q)^*$ with eigenvalues some power of q. Hence ϕ^q is a rational equivalence inducing 0 on $\tilde{H}_*(\mathbb{R}G_{u,v};\mathbb{Z}/q)$ and $\mathbb{R}G_{u,v}$ has positive weights.

Lemma 1.2 $\mathbb{R}G_{u,v}$ is simple if uv is odd.

Proof In case uv is odd it is elementary to see that the covering transformation in the 2-fold universal cover $\widetilde{\mathbb{R}G}_{u,v}$ extends to a circle action and is therefore homotopic to the identity; hence $\mathbb{R}G_{u,v}$ is simple.

Corollary 1.3 If uv is odd, then $\mathbb{R}G_{u,v}$ has positive weights.

We will need the following lemma on maps into spaces with positive weights.

Lemma 1.4 Let X be a finite complex and W a (finite nilpotent) complex with positive weights. If $\phi\colon X \to W_p$ is given, then there exists a self homotopy equivalence $\theta\colon W_p \to W_p$ such that $\theta\phi$ lifts to W:

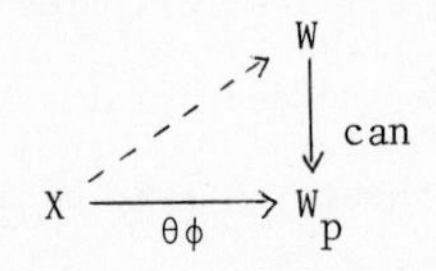

Proof By refining $W \to W_p$ into a sequence of principal fibrations with fibers $K(G,i)$'s with G torsion groups, we see that all obstructions to lifting ϕ are of finite order prime to p. Since W has positive weights we can then find a suitable p-equivalence $f : W \to W$ such that, with $\theta = f_p$, $\theta\phi$ lifts to W.

§ 2. The proof of the main results

Theorem 0.1: Using that M, N and V are $\mathbb{Z}/2\mathbb{Z}$-good this follows precisely as the main result of [GH], by replacing the Hasse principle for localization of [BK] and [HMR] with the fracture lemma for completion (cf. [BK]) which says that

$$\begin{array}{ccc} [N,W] & \to & [N,\hat{W}] \\ \downarrow & & \downarrow \\ [N,W_0] & \to & [N,(\hat{W})_0] \end{array}$$

is a pull-back diagram.

Corollary 0.2: Let $M = \Sigma^{2n+1}/H$ and $V = W = (\Sigma^{2n+2s+1}/H) \times \mathbb{R}^{2t+1}$. Note that $pr\colon \Sigma^{2n+1}/H \to \Sigma^{2n+1}/G = N$ is a $\mathbb{Z}/2$-equivalence. We define λ to be $\hat{pr}_2^{-1}$. Now M and $V = W$ are nilpotent since H is a nilpotent group acting freely and so N is $\mathbb{Z}/2\mathbb{Z}$-good. Let $j\colon M \to V$ be the obvious embedding.

Then $N \subseteq W$ by Theorem 0.1. Similarly one proves part (ii) of the Corollary.

Corollary 0.3: We apply Corollary 0.2 with $H = \{1\}$ to conclude that $\Sigma^{2n+1}/G \subseteq S^{2n+3} \times \mathbb{R}^{1+2t}$ if $2t \geq n - 2$. Hence $\Sigma^{2n+1}/G \subseteq \mathbb{R}^{2[3(n+1)/2]}$.

Corollary 0.4: This is immediate from Theorem 0.1, since a π-manifold immerses in euclidean space with codimension 1.

Theorem 0.5: Consider the composite map $\alpha : N \xrightarrow{can} N_2 \xrightarrow{\zeta} M_2 \xrightarrow{j_2} V_2 \xrightarrow{\sigma} W_2$. Since W has positive weights, we can find a $\theta: W_2 \simeq W_2$ such that $\theta\alpha$ lifts to a map $i : N \to W$ (cf. Lemma 1.4). It follows now that the hypotheses of Theorem 0.1 are fulfilled by choosing $\lambda = \hat{\zeta}_2$, $\mu = (\theta\sigma)^{\wedge}_2$ and i_0, $\hat{i}_{odd}$ the maps induced by $i : N \to W$. The diagram (iv) is then commutative by construction. Hence $N \subseteq W$ by Theorem 0.1

Corollary 0.6: For (i) we choose $M = \mathbb{R}G_{u,v} \subseteq \mathbb{R}G_{u+r,v+s} \times \mathbb{R}^W = V = W$. W has positive weights by Corollary 1.3. Since can: $\mathbb{R}G_{u,v} \to (\mathbb{R}G_{u,v}/C_{2t+1})$ is a 2-equivalence we may choose $\lambda = \hat{can}_2^{-1} : \hat{N}_2 \to \hat{M}_2$ where $N = (\mathbb{R}G_{u,v}/C_{2t+1})$. We obtain then $N \subseteq W$ by Theorem 0.5. The proof of (ii) is similar.

REFERENCES

[BK]: A.K. Bousfield and D.M. Kan, Homotopy Limits, Completions and Localizations, Springer Lecture Notes in Math. Vol. 304.

[BS]: R. Body and D. Sullivan, Homotopy types which telescope (preprint).

[F]: E. Friedlander, Maps between localized homogeneous spaces (preprint).

[GH]: H. Glover and W. Homer, Immersing manifolds and 2-equivalence (to appear in Proceedings of Northwestern University Conference on Geometrical Applications of Homotopy Theory, Springer Lecture Notes in Math.).

[HMR]: P. Hilton, G. Mislin and J. Roitberg, Localization of Nilpotent groups and spaces, North Holland, 1975.

[P]: T. Petrie, Free metacyclic group actions on homotopy spheres, Ann. of Math. 94 (1971), 108-124.

BP HOMOLOGY AND FINITE H-SPACES

RICHARD KANE

This paper is an extension of the work in [8] and [10]. It is centered around applications to finite H-spaces. Let Q_p be the integers localized at the prime p. Let $H_*(X)$ and $K_*(X)$ be ordinary and K homology, both with Q_p coefficients. Let $BP_*(X)$ be the BP homology of X. It is a module over

$$\Lambda = BP_*(pt) = Q_p[v_1,v_2,\ldots] \quad (\deg v_s = 2p^s-2).$$

Consider the paper [8]. In it we studied $BP_*(X)$ under the hypothesis that (X,μ) is a 1-connected H-space of finite type and $H_*(\Omega X)$ is torsion free. (ΩX is the loop space.) This hypothesis is motivated by finite H-space theory. For it is known that if (X,μ) is a 1-connected (mod p) finite H-space then $H_*(\Omega X)$ is torsion free for p odd (see [11]) or for $p = 2$ when X is a Lie group (see [2]). We used our results on $BP_*(X)$ to show that $K_*(X)$ is torsion free. More precisely, let $\Lambda(1) = \Lambda(\frac{1}{v_1})$ and let $BP_*(X;\Lambda(1)) = BP_*(X) \otimes_\Lambda \Lambda(1)$ be BP homology with $\Lambda(1)$ coefficients (we are localizing with respect to v_1). Then $BP_*(X;\Lambda(1))$ is torsion free (see Theorem 1:1 of [8]) and it follows that $K_*(X)$ has no torsion (see Theorem 1:4 of [8]). Thus lack of torsion in $BP_*(X;\Lambda(1))$ suffices to prove the K-theory result, which was the main point of [8]. However one can ask if a stronger BP result is true. Is $BP_*(X;\Lambda(1))$ actually a free $\Lambda(1)$ module? In particular if X is a finite complex then freeness is always obtained when we reduce mod p (see Theorem 3:1 of [5]). We will study this question for finite H-spaces. As it turns out, for p odd, $BP_*(X;\Lambda(1))$ is free.

<u>Theorem 1:1</u>. *Let* p *be odd. Let* (X,μ) *be a 1-connected (mod p) finite H-space. Then* $BP_*(X;\Lambda(1))$ *is* $\Lambda(1)$ *free.*

However, for $p = 2$, freeness fails.

<u>Theorem 1:2</u>. *For* $p = 2$, $BP_*(X;\Lambda(1))$ *is not* $\Lambda(1)$ *free if* X *is either of the exceptional Lie groups* E_7 *or* E_8.

We use an extra fact, over and above the fact that $H_*(\Omega X)$ is torsion free, in order to prove 1:1. The Milnor element Q_1 from the Steenrod algebra induces a surjective map $Q^{odd}(H^*(X;\mathbb{Z}/p)) \to Q^{even}(H^*(X;\mathbb{Z}/p))$ for p odd (see Theorem 4:4:1 of [11]). An appropriate (mod 2) analogue of the Q_1 condition holds provided (X,μ) is a 1-connected compact semi-simple Lie group prime to the exceptional Lie groups E_7 and E_8. For such spaces we can also prove $\Lambda(1)$ freeness when $p = 2$. However the condition fails for E_7 and E_8. And, as Theorem 1:2 shows, $\Lambda(1)$ freeness fails as well.

It is of interest that Lin proved the Q_1 condition on $Q(H^*(X;\mathbb{Z}/p))$ for p odd precisely to show that $K_*(X)$ has no p torsion for p odd (see [11]). Thus Theorems 1:1 and 1:2 help to clarify the relationship between his proof and that in [8] -- particularly why the arguments of [8] extend to the prime $p = 2$ as well. For the arguments in [8] never set out to prove more than the fact that $BP_*(X;\Lambda(1))$ is torsion free.

Now consider the paper [10]. In it we showed that the Chern character map, as well as its inverse, can be defined in terms of BP operations. Thus statements about the Chern character should have consequences for BP theory. In particular there is a conjecture of Atiyah and Mimura that spherical classes in $H_*(X)$ can be detected by the Chern character if X is a compact Lie group. Let

$$ch^{-1}: H_*(X) \to H_*(X) \otimes Q \xrightarrow{(ch \otimes Q)^{-1}} K_*(X) \otimes Q$$

be the inverse to the Chern character map. Let $P(H_*(X))$ be the primitive elements of $H_*(X)$ (in the sense of coalgebra primitive).

<u>Conjecture A</u> (Atiyah-Mimura). $x \in P(H_*(X))/\text{Torsion}$ *is spherical* $\Longleftrightarrow$ $ch^{-1}(x)$ *is integral i.e.* $ch^{-1}(x) \in K_*(X) \subset K_*(X) \otimes Q$.

We can relate this conjecture to a statement about BP homology. First of all we must distinguish between two types of primitivity. There is a left action of $BP^*(BP)$ on $BP_*(X)$ (see [1]). An element $x \in BP_*(X)$ will be said to be primitive if all elements of positive dimension from $BP^*(BP)$ act trivially on x. The term "primitive" will always be used in this sense. On the other hand let $P(BP_*(X))$ denote the elements which are coalgebra primitive. We will simply write $x \in P(BP_*(X))$.

onjecture B. $x \in P(BP_*(X))/\text{Torsion}$ *is spherical* $\Longleftrightarrow$ x *is primitive.*

ur reason for suggesting this conjecture is the following.

heorem 1:3. *For any space* X *Conjecture* A *implies Conjecture* B.

o, in particular A implies B when X is a compact Lie group.

The organization of this paper is as follows. In §2 we will reformulate heorems 1:1 and 1:2. In §3 we will prove 1:1. In §4 we will prove 1:2. In §5 e will prove 1:3.

Throughout this paper we will assume as known the basics of BP theory (see 4], [8], and [10]). We will also assume as known the basics of Eilenberg-Moore pectral sequence theory by which the homology or cohomology of X and ΩX are elated (see [3],[6],[7],[12]). In particular the loop map $\Omega^*\colon Q(h^*(X)) \to P(h^*(\Omega X))$ nd the delooping map $\Omega_*\colon Q(h_*(\Omega X)) \to P(h_*(X))$ will be used extensively.

2. A Criterion for $\Lambda(1)$ Freeness.

In this section we restate the problem which we are dealing with in Theorems :1 and 1:2. Assume that (X,μ) is a 1-connected (mod p) finite H-space and that $I_*(\Omega X)$ is torsion free. We will show

roposition 2:1. $BP_*(X;\Lambda(1))$ *is* $\Lambda(1)$ *free, if, and only if,* $Q(BP_*(\Omega X;\Lambda(1)))$ *is* (1) *free.*

This follows from the Eilenberg-Moore spectral sequence arguments used in [8]. We showed, at least implicitly, in the proof of 1:2(b) of [8] that the delooping map $\Omega_*\colon Q(BP_*(\Omega X;\Lambda(1)) \to P(BP_*(X;\Lambda(1)))$ is injective. So if $BP_*(X;\Lambda(1))$ is torsion free it follows that $Q(BP_*(\Omega X;\Lambda(1)))$ is $\Lambda(1)$ free. Conversely, suppose that $Q(BP_*(\Omega X;\Lambda(1)) \cong \text{Image } \Omega_*$ is $\Lambda(1)$ free. We begin by reducing mod p. Let $\Lambda_p = Z/p[v_1,v_2,\dots]$ and $\Lambda_p(1) = \Lambda_p(\frac{1}{v_1})$. In §7 of [8] we proved that

$$\text{Tor}^{BP_*(\Omega X;\Lambda_p(1))}(\Lambda_p(1);\Lambda_p(1))$$

is an exterior algebra generated by elements with external degree 1. It follows that the spectral sequence converging to $BP_*(X;\Lambda_p(1))$ collapses and that $E^0(BP_*(X;\Lambda_p(1)))$ is an exterior algebra generated by Image Ω_*. Let $\chi = \{x_i\}$ be a $\Lambda(1)$ basis of Image $\Omega_* \subset BP_*(X;\Lambda(1))$. Let $M(\chi)$ be the monomials $\{x_1^{s_1}\ldots x_n^{s_n} | s_i = 0,1\}$. Since $BP_*(X;\Lambda(1))$ is torsion free it follows from the above that $M(\chi)$ generates $BP_*(X;\Lambda(1))$ as a $\Lambda(1)$ module. Furthermore, by standard coalgebra arguments, the fact that there are no $\Lambda(1)$ relations between the elements of χ implies that the same is true for the elements of $M(\chi)$. It follows that $BP_*(X;\Lambda(1))$ is a free $\Lambda(1)$ module.

§3. Proof of Theorem 1:1.

In this section we prove Theorem 1:1. Assume for the rest of this section that p is an odd prime. Because of 2:1 it suffices to show that $Q(BP_*(\Omega X;\Lambda(1)))$ is not $\Lambda(1)$ free. We will do so by utilizing information about $H^*(X;\mathbb{Z}/p)$. Our proof will be divided into two parts. In the first part we will explain how $H^*(X;\mathbb{Z}/p)$ is related to $Q(BP_*(\Omega X;\Lambda(1)))$. In the second part we will establish the precise technical results needed to prove 1:1.

PART I. The algebra structure of $H_*(\Omega X;\mathbb{Z}/p)$ is related to the Steenrod module structure of $Q(H^*(X;\mathbb{Z}/p))$ via an Eilenberg-Moore spectral sequence. In particular

(3:1) $H_*(\Omega X;Z/p) = Z/p[X]/I \otimes Z/p[Y]$ where I is the ideal generated by $\{x_i^p | x_i \in X\}$.

This is deduced from the fact that

(3:2) There exist elements $\{a_i\}$ in $Q^{odd}(H^*(X;\mathbb{Z}/p))$ such that

(a) $\{\beta_p P^{m_i}(a_i)\}$ is a basis of $Q^{even}(H^*(X;\mathbb{Z}/p))$ $(|a_i| = 2n_i+1)$

(b) $\langle\Omega^*(a_i),x_j\rangle = \delta_{ij}$ (th Kronecker delta).

The point is that the differential d_{p-1} in the spectral sequence can be identified with the Steenrod operation $\beta_p P^m$ acting in $Q(H^*(X;\mathbb{Z}/p))$ and it is this differential acting nontrivially which produces elements truncated at height p (see [6] for more details).

Since $H_*(\Omega X)$ is torsion free we have surjective maps

$$BP_*(\Omega X) \xrightarrow{T} H_*(\Omega X) \xrightarrow{\rho} H_*(\Omega X;\mathbf{Z}/p).$$

Here T is the Thom map and ρ is reduction mod p.) Let $\hat{Z} = \hat{X} \cup \hat{Y}$ be representatives in $BP_*(\Omega X)$ for the elements $Z = X \cup Y$. Let D be the set of monomials in the elements of $\hat{Z}$ which do not include the pth power of any element from $\hat{X}$. Then $\hat{Z} \cup D$ is a Λ basis of $BP_*(\Omega X)$. In fact

(3:3) $BP_*(\Omega X)$ is isomorphic, as an algebra, to $\Lambda[\hat{Z}]/J$ where J is the ideal generated by the elements $\{R_X | X \in \hat{X}\}$ where each R_X is of the form

$$R_X = X^p - \sum \lambda_i Z_i - \sum \omega_j d_j$$

where $Z_i \in \hat{Z}$, $d_j \in D$, and $\lambda_i, \omega_j \in \Lambda$.

Thus J defines the relations by which monomials in $\hat{Z}$ involving p^{th} powers of elements from $\hat{X}$ can be written in terms of $\hat{Z} \cup D$. It follows from 3.3 that

(3:4) $Q(BP_*(\Omega X))$ is isomorphic, as a Λ module, to the quotient M/L where M is the free Λ module on generators $\hat{Z}$ and L is the submodule generated by the relations $\{Q_X\}$ where Q_X is determined from R_X of 3:3 by the rule

$$Q_X = \sum \lambda_i Z_i.$$

The relations in 3:3 and 3:4 can be partially determined from a knowledge of the Steenrod module structure of $H_*(\Omega X;\mathbf{Z}/p)$. This follows from the next two facts. Let r_1 be the Quillen operation acting on $BP_*(\Omega X)$. It is related to the Steenrod power P^1 by the following commutative diagram (see 2:4 of [8]).

(3:5)
$$\begin{array}{ccc} BP_*(\Omega X) & \xrightarrow{r_1} & BP_*(\Omega X) \\ \downarrow \rho T & & \downarrow \rho T \\ H_*(\Omega X;\mathbf{Z}/p) & \xrightarrow{-P^1} & H_*(\Omega X;\mathbf{Z}/p) \end{array}$$

Let $(\Omega\Delta)^n: BP_*(\Omega X) \to \bigotimes_{i=1}^{n} BP_*(\Omega X)$ be the reduced comultiplication.

(3:6) For each $X \in \hat{X}$, $X^p \equiv pU - v_1 r_1(U) + v_1 d$ modulo elements of lower filtration where $(\Omega\Delta)^P(U) \equiv X \otimes \cdots \otimes X$ modulo elements of lower filtration and d is decomposable.

(Here the filtration is the skeleton filtration. See 5:2 of [8] for the above.) Finally we observe that we can pass from $BP_*(\Omega X)$ to $BP_*(\Omega X;\Lambda(1))$ by tensoring by $\Lambda(1)$ and 3:3, 3:4, and 3:6 will still be valid.

PART II. Let us begin with $Q(BP_*(\Omega X;\Lambda(1)))$. By 3:4 it suffices to prove that the elements $\{Q_{X_i}\}$ are part of a $\Lambda(1)$ basis of M. Thus, by 3:6, it suffices to prove that the elements $\{r_1(U_i)\}$ are part of a $\Lambda(1)$ basis of M. Passing to $Q(BP_*(\Omega X))$ it suffices to show that the elements $\{r_1(U_i)\}$ are part of a Λ basis of M. We have surjective maps $M \to Q(BP_*(\Omega X)) \to Q(H_*(\Omega X;\mathbb{Z}/p))$. Since the Λ basis of M given by 3:4 projects to a $\mathbb{Z}/p$ basis of $Q(H_*(\Omega X;\mathbb{Z}/p))$ it follows that any set of elements which project to a basis of $Q(H_*(\Omega X;\mathbb{Z}/p))$ must be a basis of M. Thus it suffices to prove that the elements $\{r_1(U_i)\}$ project to a linearly independent set in $Q(H_*(\Omega X;\mathbb{Z}/p))$. Let $u_i = \rho T(U_i)$ and let $w_i = \Omega_*(u)$. By 3:5 it suffices to prove

(3:7) $\{P^1(w_i)\}$ is a linearly independent set in $P(H_*(X;\mathbb{Z}/p))$.

We can prove 3:7 by proving the following three lemmas. Let

$$K = Q(H^*(X;\mathbb{Z}/p)) \cap \text{kernel } \beta_p.$$

Lemma 3:8. $\langle K,w_i\rangle = 0$ *for each* w_i.

Thus there is a well defined pairing between the elements w_i and the elements in $Q = Q(H^*(X;\mathbb{Z}/p))/K$. With respect to this pairing

Lemma 3:9. Q *has a basis* $\{b_i\}$ *such that* $\langle b_i,w_j\rangle = \delta_{ij}$.

Also

Lemma 3:10. *The map* $Q(H^*(X;\mathbb{Z}/p)) \xrightarrow{P^1} Q(H^*(X;\mathbb{Z}/p)) \to Q$ *is surjective.*

Property 3:7 will follow. For, by 3:10, the basis $\{b_i\}$ in 3:9 can be chosen to be of the form $\{P^1(c_i)\}$. Then

$$\langle c_i,P^1(w_j)\rangle = \langle P^1(c_i),w_j\rangle = \delta_{ij}$$

and $\{P^1(w_j)\}$ must be linearly independent. Hence we are left with proving the above lemmas.

Proof of Lemma 3:8. First of all, since $H^*(X)$ has no higher p torsion (see 1:4 of [6]) and the image of $\beta_p: Q(H^*(X;Z/p)) \to Q(H^*(X;Z/p))$ is K, it follows from a Bockstein spectral sequence argument that the map $\rho: Q(H^*(X)) \to Q(H^*(X;Z/p))$ maps onto K.

Now pick $k \in K$. Thus $k = \rho(\ell)$ for some $\ell \in Q(H^*(X))$ and

$$
\begin{aligned}
\langle k, w_i\rangle &= \langle \rho(\ell), \Omega_*(u_i)\rangle \\
&= \langle \rho\Omega^*(\ell), u_i\rangle \\
&= \langle \rho\Omega^*(\ell), \ell T(U_i)\rangle \\
&= \rho\langle \Omega^*(\ell), T(U_i)\rangle .
\end{aligned}
$$

Thus, to show $\langle k, \Omega_*(w_i)\rangle = 0$, it suffices to show that $\langle \Omega^*(\ell), T(U_i)\rangle = 0$. But $\langle \Omega^*(\ell), T(U_i)\rangle \in Q_p$. Hence it suffices to show that $\rho\langle \Omega^*(\ell), T(U_i)\rangle = 0$

$$
\begin{aligned}
\rho\langle \Omega^*(\ell), T(U_i)\rangle &= \langle \Omega^*(\ell), pT(U_i)\rangle \\
&= \langle \Omega^*(\ell), T(X_i)^p\rangle \\
&= 0.
\end{aligned}
$$

The second equality follows from 3:6. The third equality follows from the fact that $\Omega^*(\ell)$ is primitive while $T(X_i)^p$ is decomposable.

Proof of Lemma 3:9. Letting a_i be as in 3:2 we will prove 3:9 by letting $b_i = P^{m_i}(a_i)$. By 3:2(a), $\{b_i\}$ projects to a basis of Q. Letting $c_i = \Omega_*(a_i)$ and $(\Omega\Delta)_n: H_*(\Omega X;Z/p) \to \bigotimes_{i=1}^{n} H_*(\Omega X;Z/p)$ be the n fold reduced comultiplication we have the following:

$$
\begin{aligned}
\langle c_i, w_j\rangle &= \langle P^{m_i}(a_i), \Omega_*(u_j)\rangle \\
&= \langle P^{m_i}(c_i), u_j\rangle \\
&= \langle c_i^p, u_j\rangle \\
&= \langle c_i \otimes \cdots \otimes c_i, (\Omega\Delta)_p(u_j)\rangle \\
&= \langle c_i \otimes \cdots \otimes c_i, x_j \otimes \cdots \otimes x_j\rangle \\
&= \delta_{ij}.
\end{aligned}
$$

The third equality comes from the fact that $|c_i| = 2m_i$. The last two equalities come from 3:6 and 3:2(b).

Proof of Lemma 3:10. The Milnor element $Q_1 = \beta_p P^1 - P^1\beta_p$ induces a surjective map

$$Q_1 : Q^{odd}(H^*(X;\mathbf{Z}/p)) \to Q^{even}(H^*(X;\mathbf{Z}/p))$$

(see 4:4:1 of [11]). Also P^1 acts trivially on $Q^{even}(H^*(X/\mathbf{Z}/p))$ since $Q^{2n}(H^*(X;\mathbf{Z}/p)) = 0$ unless $n \equiv 1 \pmod p$ (see 4:3:1 of [11]). Thus the map

$$\beta_p P^1 : Q^{odd}(H^*(X;\mathbf{Z}/p)) \to Q^{even}(H^*(X;\mathbf{Z}/p))$$

is surjective. This suffices to establish 3:10.

§4. Proof of Theorem 1:2.

In this section we prove Theorem 1:2. Assume for the rest of this section that we are dealing with the prime $p = 2$. We will concentrate on the case $X = E_8$ since the case $X = E_7$ is similar but simpler. It will be dealt with briefly at the end of this section.

By 2:1 we need only show that $Q(BP_*(\Omega E_8;\Lambda(1)))$ is not $\Lambda(1)$ free. Our approach is analogous to that employed in the previous section. We will divide our proof into two parts. In part I we will use $H^*(E_8;\mathbf{Z}/2)$ and $H_*(\Omega E_8;\mathbf{Z}/2)$ to study $BP_*(\Omega E_8)$. In part II we will study $Q(BP_*(\Omega E_8;\Lambda(1)))$ and prove lack of freeness.

PART I. Recall the following facts about $H_*(\Omega E_8;\mathbf{Z}/2)$.

(4:1) $H_*(\Omega E_8;Z/2)$ is isomorphic, as an algebra, to $T \otimes P$ where

$$T = \mathbf{Z}/2[x_2,x_4,x_8,x_{14}]/\langle x_2^2,x_4^2,x_8^2,x_{14}^2\rangle$$

$$P = \mathbf{Z}/2[x_{16},x_{22},x_{26},x_{28},x_{34},x_{38},x_{46},x_{58}].$$

(Here x_s is an element of dimension s.)

(4:2) $Sq^2(x_4) = x_2, Sq^2(x_{16}) = x_{14}, Sq^2(x_{28}) = x_{26},$

$Sq^{02}(x_8) = x_2.$

These facts can be deduced from the structure of $H^*(E_8;\mathbf{Z}/2)$ as an algebra over the Steenrod algebra (see [13]). The reinterpretation of these results in the form of

4:1 and 4:2 is obtained by using an Eilenberg-Moore spectral sequence (see [6] and 7:1 of [12]).

From 4:1 and 4:2 we can deduce all that we need to know about the algebra structure of $BP_*(\Omega E_8)$. Pick representatives $\hat{X} = \{X_s\}$ in $BP_*(\Omega E_8)$ for the elements $X = \{x_s\}$ in $H_*(\Omega E_8;\mathbb{Z}/2)$. From 4:1 we can deduce

(4:3) $BP_*(\Omega E_8)$ is isomorphic, as an algebra, to $\Lambda[\hat{X}]/J$ where J is the ideal generated by elements

$$R_4 = X_2^2 - \Sigma\alpha_s X_s + d_4$$

$$R_8 = X_4^2 - \Sigma\beta_s X_s + d_8$$

$$R_{16} = X_8^2 - \Sigma\gamma_s X_s + d_{16}$$

$$R_{28} = X_{14}^2 - \Sigma\Delta_s X_s + d_{28}$$

where the elements $\{\alpha_s,\beta_s,\gamma_s,\Delta_s\}$ are from Λ while d_4, d_8, d_{16}, and d_{28} are decomposable.

By using 4:2 we can put restrictions on the coefficients in R_4, R_8, R_{16} and R_{28}.

(4:4) Up to units in Q_2 we have the following identities:

(a) $\alpha_4 = \beta_8 = \gamma_{16} = \Delta_{28} = 2$

(b) $\alpha_2 = \gamma_{14} = \Delta_{26} = v_1$

(c) $\beta_4 = v_1^2$.

The identities (a) and (b) follow from the mod 2 analogues of 3:5 and 3:6. Identity (c) requires a more involved argument. See 6:2 of [8].

Remark. When we employ 4:4 the precise value of the units which should appear in (a), (b), and (c) is immaterial. Thus, from now on, we will assume that they are all equal to 1 and, hence, that (a), (b), and (c) are true identities.

We can also obtain partial information on the coefficient β_2 in R_8. For dimension reasons β_2 is a linear combination of v_2 and v_1^3

$$\beta_2 = av_2 + bv_1^3 .$$

(4:5) a is an unit in Q_2.

This follows by arguments analogous to those used in [8] to prove 3:5 and 3:6 of this paper. We repeat the proof of 5:2 of [8] only replacing the use of the operation r_1 by the operation r_{01}. Since $Sq^{02}(x_8) = x_2$ the lemma follows from 2:4 of [8] which establishes a correspondence between r_{01} and Sq^{02}.

PART II. It follows from 4:3 that

(4:6) $Q(BP_*(\Omega E_8))$ is isomorphic, as a Λ module, to M/L where F is the free Λ module generated by $\hat{X}$ and L is the submodule generated by Q_4, Q_8, Q_{16}, and Q_{28} where $Q_4 = \Sigma\alpha_s X_s$, $Q_8 = \Sigma\beta_s X_s$, $Q_{16} = \Sigma\gamma_s X_s$, and $Q_{28} = \Sigma\Delta_s X_s$.

Since $Q(BP_*(\Omega E_8;\Lambda(1))) = Q(BP_*(\Omega E_8)) \otimes_\Lambda \Lambda(1)$, the description in 4:6 also holds for $Q(BP_*(\Omega E_8;\Lambda(1)))$. We now modify the description so as to see that $\Lambda(1)$ freeness does not hold.

First of all, since v_1 is invertible, it follows from 4:4(b) that we can use Q_{16} and Q_{28} to express X_{14} and X_{26} in terms of the remaining elements of $\hat{X}$. Thus $Q(BP_*(\Omega E_8;\Lambda(1)))$ is isomorphic, as a $\Lambda(1)$ module, to M'/L' where M' is the free $\Lambda(1)$ module generated by $\hat{X} - \{X_{14},X_{26}\}$ and L' is the submodule generated by Q_4 and Q_8.

Similarly we can eliminate X_2 in Q_4. However, since X_2 also appears in Q_8 this elimination involves the rewriting of Q_8. By 4:4 and 4:5 the relations Q_4 and Q_8 can be written as

$$Q_4\colon\ 2X_4 + v_1X_2 = 0$$

$$Q_8\colon\ 2X_8 + v_1^2X_4 + (av_2 + bv_1^3)X_2 = 0.$$

Rewriting Q_4 as $X_2 = -2v_1^2X_4$ the equations reduce to

$$Q_8'\colon\ 2X_8 + \beta_4'X_4 = 0$$

where $-\beta_4' = (2b-1)v_1^2 + 2av_1^{-1}v_2$. The fact that $Q(BP_*(\Omega E_8;\Lambda(1)))$ is not $\Lambda(1)$ free will follow if we can show that β_4' is not invertible in $\Lambda(1)$. Since the coefficients of both v_1^2 and $v_1^{-1}v_2$ in β_4' are nonzero we need only show

Lemma 4:7. *Up to units in* Q_2 *the only invertible elements in* $\Lambda(1)$ *are the elements* $\{v_1^n\}$ *for* $n \in \mathbb{Z}$.

Proof: All elements in $\Lambda(1)$ can be written $\frac{x}{v_1^s}$ where $s \geq 0$ and $x \in \Lambda$. If $\frac{x}{v_1^s} \cdot \frac{y}{v_1^t} = 1$ then $xy = v_1^{s+t}$. Since Λ is a polynomial algebra it follows that, up to units in Q_2, x and y are powers of v_1. Q.E.D.

Remark. The case $X = E_7$ is similar but simpler. The argument to prove 1:2 for $X = E_7$ amounts to going through the argument for $X = E_8$ but ignoring the relation Q_{28}.

§5. Proof of Theorem 1:3.

In this section we prove Theorem 1:3. Let X be a space and suppose that conjecture A is true for X. In showing conjecture B is true it is trivial that x spherical implies x primitive. So suppose, conversely, that x is primitive. We have an embedding

$$BP_*(X)/\text{Torsion} \hookrightarrow BP_*(X) \otimes Q \tag{5:1}$$

(the point is that all torsion in $BP_*(X)$ i.e. v_1 torsion, v_2 torsion, etc., is also p torsion (see [4])). In [10] we defined an operation

$$P: BP_*(X) \otimes Q \to BP_*(X) \otimes Q$$

which has the property of characterizing primitive elements. That is, x is primitive if, and only if, $x = P(y)$ for some $y \in BP_*(X) \otimes Q$. Also $P^2 = P$. Therefore, we can also say that x is primitive if, and only if $P(x) = x$. The map P factors uniquely through the Thom map

$$\begin{array}{ccc} BP_*(X) \otimes Q & \xrightarrow{P} & BP_*(X) \otimes Q \\ \downarrow & \nearrow_{\overline{P}} & \\ H_*(X) \otimes Q & & \end{array} \tag{5:2}$$

Also, if we follow $\overline{P}$ by the Conner-Floyd map then we obtain the inverse to the Chern character map

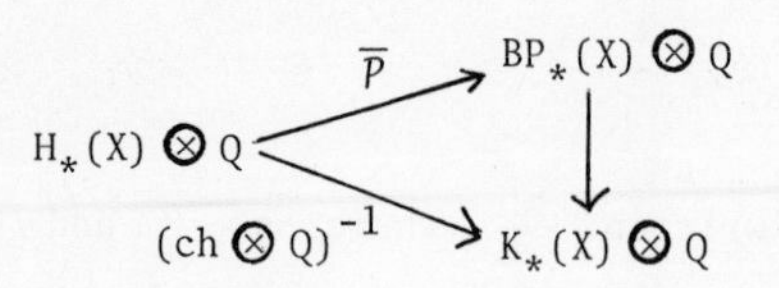

There is a well defined map $P(BP_*(X))/\text{Torsion} \to P(H_*(X))/\text{Torsion}$ (see the remark after 5:1). Let y be the image of x in $P(H_*(X))/\text{Torsion}$. By the above $ch^{-1}(y)$ is the image of x under the map $BP_*(X)/\text{Torsion} \to BP_*(X) \otimes Q \xrightarrow{P} BP_*(X) \otimes Q \to$ $\to K_*(X) \otimes Q$. Since $P(x) = x \in BP_*(X) \subset BP_*(X) \otimes Q$ it follows that

$$ch^{-1}(y) \in K_*(X) \otimes Q.$$

Then, by conjecture A, y is spherical. By the commutative diagram

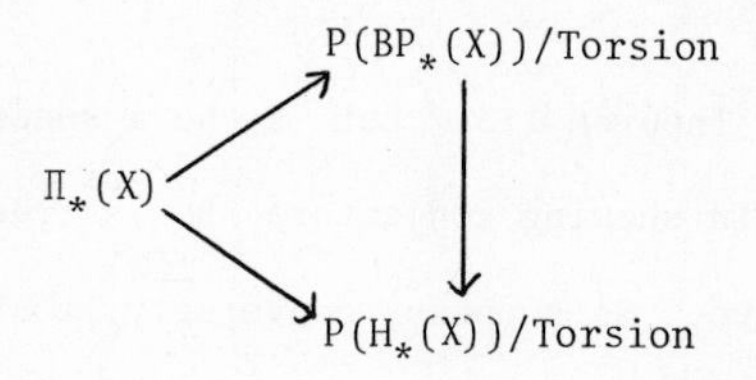

it follows that y has a representative z in $P(BP_*(X))/\text{Torsion}$ which is spherical. But $x = z$. Consider $P(BP_*(X))/\text{Torsion}$ as imbedded in $BP_*(X) \otimes Q$. By 5:2 $P(x) = P(z)$ $(= \overline{P}(y))$. Also, x and z are primitive. Thus $P(x) = x$ and $P(z) = z$.

REFERENCES

[1] Adams, J.F. Lectures on Generalized Cohomology, Lecture Notes in Mathematics, 99, Springer-Verlag (1969).

[2] Bott, R. The Space of Loops on a Lie Group, Mich. J. Math. (1958), 35-61.

[3] Clark, A. Homotopy Commutativity and the Moore Spectral Sequence, Pacific J. Math. 15 (1965), 65-74.

[4] Johnson, D.C., and Wilson, W.S. Projective Dimension and Brown-Peterson Homology, Topology 12 (1973), 327-353.

[5] Johnson, D.C., and Wilson, W.S. BP Operations and Morava's Extraordinary K-Theories, Math. Z. 144 (1975), 55-75.

[6] Kane, R. On Loop Spaces without p Torsion, Pacific J. Math. 60 (1975), 189-201.

[7] Kane, R. On Loop Spaces without p Torsion II, Pacific J. Math. (to appear).

[8] Kane, R. The BP Homology of H-Spaces, Trans. Amer. Math. Soc. (to appear).

[9] Kane, R. On Spherical Homology Classes, Quart, J. Math. Oxford (to appear).

[10] Kane, R. BP Operations and the Chern Character (to appear).

[11] Lin, J. Torsion in H-Spaces II, Annals of Math (to appear).

[12] Petrie, T. The Weakly Complex Bordism of Lie Groups, Annals of Math. 88 (1968), 370-402.

[13] Thomas, E. Exceptional Lie Groups and Steenrod Squares, Mich. J. Math. 11 (1964), 151-156.

ON IMMERSIONS $CP^n \looparrowright R^{4n-2\alpha(n)}$

François Sigrist and Ueli Suter

(with the collaboration of P. J. Erard)

§1. Introduction

In this paper, we shall show how complex K-theory can provide non-immersion results for CP^n. Immersions (non-immersions) will be noted $\looparrowright$ ($\not\looparrowright$), embeddings (non-embeddings) $\hookrightarrow$ ($\not\hookrightarrow$).

A considerable amount of information on this subject is contained in a survey article by I. M. James: Euclidean models of projective spaces [5], as well as a very extensive bibliography. For this reason, we have reduced our references to a minimum, urging the reader to keep James' paper at hand.

The function $\alpha(n)$ (= number of ones in the dyadic expansion of n) plays a central role in our problem. To illustrate this, let us give four results which have been for us a motivation and a starting point:

Theorem A: $CP^n \hookrightarrow R^{4n-\alpha(n)}$ (Steer [8])

Theorem B: $CP^n \not\hookrightarrow R^{4n-2\alpha(n)}$ (Atiyah-Hirzebruch [3])

Theorem C: $CP^n \not\looparrowright R^{4n-2\alpha(n)-1}$ (Sanderson-Schwarzenberger, Mayer [7, 6])

Theorem D: $CP^n \looparrowright R^{4n-2\alpha(n)}$ if $n=2^s+3$ (Steer [9])

We shall stick to the situation $CP^n \looparrowright R^{4n-2\alpha(n)}$, and give necessary conditions for n. Most, if not all, results on this problem deal with n even. Our investigation applies to any n: it turns out that our conditions cross the known results in a very surprising way. To express our main result, we shall use the Stirling numbers of the first kind, familiar to topologists since their generating function is

$$\sum_{q\geq 0} \frac{m!}{(m+q)!} S(m+q,m)t^q = \left(\frac{\log(1+t)}{t}\right)^m$$

We then have:

<u>Theorem 3:</u> Suppose $CP^n \looparrowright R^{4n-2\alpha(n)}$. Then there exists an integer e_o such that $\frac{e_o \cdot (2n-\alpha(n)+1)!}{(2n-\alpha(n)+1+k)!} S(2n-\alpha(n)+1+k,\ 2n-\alpha(n)+1)$ is - even for $0\leq k\leq \alpha(n)-2$ - odd for

$k = \alpha(n)-1$ and $k = \alpha(n)$.

We have been unable to formulate these conditions, in their full strength, in a more palatable way. But to get an idea of the coverage obtained, we had the luck of knowing a competent (and crafty) computer scientist, our colleague P. J. Erard, to whom we express our warmest thanks. Taking advantage of the fact that Stirling numbers of the first kind satisfy a Pascal-type relation $S(m,r) = S(m-1,r-1) - (m-1)\, S(m-1,r)$, Erard could check the conditions of theorem 3 for $n \leq 2^{14}$, a reasonably good sample. The distribution of favourable values of n offers numerous regularities; however, we could not extract any "good guess". On the other hand, from the purely quantitative point of view, the elimination rates obtained in various sub-samples deserve mention. We shall devote §4 to some computer records. The main fact is that only 4106 values among 16384 satisfy the conditions of theorem 3; the elimination rate is therefore very close to 75%. Also, a very pleasant surprise is furnished, to take a specific example, by the family $\{n \text{ even}, \alpha(n) \equiv 0 (\text{mod } 4)\}$, in which, according to [5], no non-immersion is known. Our sample contains 2015 values, among which only 178 satisfy the conditions, an elimination exceeding 90%.

A subsequent paragraph deals with low values of $\alpha(n)$, for which the conditions of theorem 3 can be explicitely given. The first relevant result is that the wanted immersion is impossible if $\alpha(n) = 2$. This is proved by theorem 3, except if $n = 3$, where a separate argument is needed. Next, for $\alpha(n) = 3$, theorem 3 eliminates all values of n, except precisely those for which the immersion is constructed by Steer (theorem D). We finally include $\alpha(n) = 4$ and 5.

In §6, we first prove that the conditions of theorem 3 imply that the binomial coefficient $\binom{n+1}{\alpha(n)}$ is even. This condition is much weaker, but allows to produce many infinite families of non-immersion values. It suffices in particular to exhibit infinitely many values of $n \equiv 3 \pmod 4$ without immersion, disproving a rather current conjecture. It should be noted that our numerical results on $n \equiv 3 \pmod 4$ give a very low elimination rate (36%), in sharp contrast with the other classes of congruence mod 4. We further show that non-immersion is guaranteed as soon as n is divisible by a high power of 2, a hint towards the fact that a more detailed information on the dyadic expansion of n is probably necessary to control the immersion problem.

§2. Homotopy Theory and K-Theory

We denote by $2+\beta$ the Hopf bundle over CP^n, $\beta\varepsilon\widetilde{KO}(CP^n)$. The tangent bundle $\tau(CP^n)$ satisfies the well-known equation $\tau+2 = (n+1)(2+\beta)$. If $CP^n \hookrightarrow R^{2n+k}$, the normal bundle to the immersion is consequently identified in $KO(CP^n)$: $\nu = k-(n+1)\beta$. Using $\cong$ to denote fiber homotopy equivalence, let us choose an integer b satisfying $2b \cong b(2+\beta)$. This requires that b is a multiple of the James number b_{n+1}, exponent of the finite group $J(CP^n)$. We further require b to be divisible by a suitably high power of 2, which we shall adjust later according to our needs. We now have:

$\nu+2b\cong\nu+b(2+\beta) = k+2b+(b-n-1)\beta = k+2n+2+(b-n-1)(2+\beta)$. The number b is evidently big enough to authorize cancellation, so we get the fundamental equation

$$\text{(I)} \qquad \nu+(2b-k-2n-2) \cong (b-n-1)(2+\beta)$$

This equation tells us that ν is stably fiber homotopy equivalent to a positive multiple of the Hopf bundle. As a consequence, we can explicitely identify the stable homotopy type of the Thom complex $T[\nu]$, using Atiyah's results [2]:

$$\text{(II)} \qquad \Sigma^{2b-k-2n-2}T[\nu] \simeq CP^{b-1}/CP^{b-n-2}$$

We begin by studying the suspension properties of K-theory in this situation. The necessary data are provided by:

Theorem 1:

(i) $\widetilde{KU}(CP^{b-1}/CP^{b-n-2})$ is free abelian on n+1 generators $\gamma_i (0\leq i\leq n)$ of respective filtration $2b-2n-2+2i$.

(ii) The Adams operation ψ^2 satisfies

$$\left.\begin{aligned} \psi^2(\gamma_o) &\equiv 2^{b-2n+\alpha(n)-1}(\gamma_{n-1}+\gamma_n) \\ \psi^2(\gamma_i) &\equiv 0 \end{aligned}\right\} \bmod 2^{b-2n+\alpha(n)}$$

Proof: (i) is clear, using restriction from $KU(CP^{b-1})$. Similarly, one gets $\psi^2(\gamma_i) = \sum_{q=0}^{n-i} 2^{b-n-1+i-q}\binom{b-n-1+i}{q}\gamma_{i+q}$. To get the congruences of (ii), let us first adjust the power of 2 dividing b to guarantee $\alpha(b-c)=\alpha(b-1)-\alpha(c-1)$ for $c\leq n$.

Using $\sim$ to denote divisibility of rational numbers by the same power of 2, we successively have:

$$2^{b-n-1+i-q}\binom{b-n-1+i}{q} \sim 2^{b-2n+\alpha(n)-1}\cdot 2^{n-\alpha(n)}\cdot 2^{i-q}\binom{n-i+q}{q} \sim$$

$$2^{b-2n+\alpha(n)-1}\cdot n!\cdot\frac{(2i)!}{i!}\cdot\frac{q!}{(2q)!}\binom{n-i+q}{q} \sim 2^{b-2n+\alpha(n)-1}\cdot(2i)!\,\frac{(n-i+q)!}{(2q)!}\binom{n}{i}$$

As $q \leq n-i$, the wanted congruences are evident at this stage. This completes the proof.

Theorem 1 enables us to reprove theorems B and C of §1, with a crucial additional fact:

<u>Theorem 2:</u>

(i) $CP^n \not\hookrightarrow R^{4n-2\alpha(n)-1}$

(ii) $CP^n \not\subset R^{4n-2\alpha(n)}$

(iii) If $CP^n \hookrightarrow R^{4n-2\alpha(n)}$, the multiplicative structure of $T[\nu]$ is non-trivial in KU-theory, and therefore also in integral cohomology.

<u>Proof:</u> We use the following facts in KU-theory: the Adams operation ψ^2 is divisible by 2 in $\widetilde{KU}(X)$ if either a) X is a double suspension, or b) the multiplicative structure of $\widetilde{KU}(X)$ is trivial. Fact a) comes from the Bott periodicity theorem, and fact b) is "the oldest trick of K-theory": $\psi^p(\alpha) \equiv \alpha^p (\text{mod } p)$. Now, to prove (i), deny, and deduce from equation (II):

$$\Sigma^{2(b-2n+\alpha(n)-1)}(\Sigma T[\nu]) \simeq CP^{b-1}/CP^{b-n-2}$$

By a) and b), $\psi^2(\gamma_o)$ must be divisible by $2^{b-2n+\alpha(n)}$, but this contradicts theorem 1(ii), q.e.d. Now suppose $CP^n \hookrightarrow R^{4n-2\alpha(n)}$. From equation (II) again, we get

$$\Sigma^{2(b-2n+\alpha(n)-1)}(T[\nu]) \simeq CP^{b-1}/CP^{b-n-2}$$

This shows that the multiplicative structure of $\widetilde{KU}(T[\nu])$ cannot be trivial, by the same argument, and this proves (iii). But (ii) is simultaneously proved, because cup products are trivial in $T[\nu]$ in the case of an embedding, the Euler class being zero. This completes the proof.

§3. Multiplicative structure of $T[\nu]$, and main result:

From now on we suppose $CP^n \hookrightarrow R^{4n-2\alpha(n)}$. By theorem 2, the multiplicative structure of $H^*(T[\nu])$ is non-trivial, and is therefore completely controlled by a

"Euler class" $e_o \epsilon Z$. Let us christen $z^{n-\alpha(n)+i}$ the cohomology generators ($0 \leq i \leq n$), obtained by Thom isomorphism. We then have: $z^q \cup z^r = e_o z^{q+r}$ as description of the cup-product in $T[\nu]$. We denote by $\mu_i (0 \leq i \leq n)$ the K-theory generators of $T[\nu]$, obtained in desuspending the generators γ_i of theorem 1. What we want to determine are the coefficients a_k^{ij} in

$$\mu_i \mu_j = \sum_k a_k^{ij} \mu_{i+j+n-\alpha(n)+k}$$

Equation (I), together with the naturality properties of Thom classes, tells us that the generators $z^{n-\alpha(n)+i}$ suspend to $y^{b-n-1+i}$ in the cohomology of $CP^{b-1}/CP^{b-n-2} (y \epsilon H^2(CP^\infty))$. We now use the Chern character, which is multiplicative and invariant by suspension.

From $ch\gamma_i = (e^y - 1)^{b-n-1+i}$ we get in desuspending

$$ch\mu_i = \left(\frac{e^z-1}{z}\right)^{b-2n+\alpha(n)-1} (e^z-1)^{n-\alpha(n)+i}$$

We then compare

$$ch\mu_i \cup ch\mu_j = e_o \left(\frac{e^z-1}{z}\right)^{2b-4n+2\alpha(n)-2} (e^z-1)^{2n-2\alpha(n)+i+j} \quad \text{and}$$

$$ch\mu_i \mu_j = \left(\frac{e^z-1}{z}\right)^{b-2n+\alpha(n)-1} \sum_k a_k^{ij} (e^z-1)^{2n-2\alpha(n)+i+j+k}$$

This very easily reduces to

(III)
$$e_o \left(\frac{t}{\log(1+t)}\right)^{b-2n+\alpha(n)-1} = \sum_k a_k^{ij} t^k$$

and this solves our problem. An immediate consequence is that the coefficients a_k^{ij} are independent of i and j (as they should be), so we simply shall write $a_k^{ij} = a_k$.

In what follows, we shall be interested only in the parity of the integers a_k for $0 \leq k \leq \alpha(n)$. This parity will not be modified if, in equation (III), we put $b = 0$. This is seen as follows: the relevant coefficients of the power series $(\log(1+t)/t)^{b_{n+1}}$ are integers [1,4], and b is a multiple of b_{n+1}; in choosing b divisible by a sufficiently high power of 2, as we still can do, we evidently ensure conservation of parity. By an abuse of notation which is harmless for our purposes, we therefore can suppose

(IV)
$$\mu_i\mu_j = \Sigma a_k \mu_{i+j+n-\alpha(n)+k}$$

$$\Sigma a_k t^k = e_o\left(\frac{\log(1+t)}{t}\right)^{2n-\alpha(n)+1}$$

We now apply the oldest trick of K-theory, in the form $\psi^2(\mu_o) \equiv \mu_o^2$ (mod 2). (The reader may feel the need of applying the trick to all generators μ_i, but this rapidly turns out to be redundant). From theorem 1 we have $\psi^2(\mu_o) \equiv \mu_{n-1} + \mu_n$ (mod 2), and so we obtain our main result:

Theorem 3: If $CP^n \hookrightarrow R^{4n-2\alpha(n)}$, there exists an integer e_o such that the power series

$$\Sigma a_k t^k = e_o\left(\frac{\log(1+t)}{t}\right)^{2n-\alpha(n)+1}$$

satisfies $a_o \equiv a_1 \equiv \cdots \equiv a_{\alpha(n)-2} \equiv 0 (\text{mod } 2)$ and $a_{\alpha(n)-1} \equiv a_{\alpha(n)} \equiv 1(\text{mod } 2)$.

§4. Some computer records on theorem 3.

Needless to say, the control of the conditions of theorem 3, as it stands, requires a prohibitive amount of computations, if it is performed by hand. We are very grateful to our colleague P. J. Erard, who established a computer program, and then checked the conditions for all $n \leq 16384$. Moreover, his very careful programming has provided us with all possible and imaginable displays and reorderings; he therefore should not share responsibility in the fact that we could not extract any reasonable guess. However, we think some samples give good indications on the strength of our theorem.

The four columns of the table below are:

N = sample considered, always taken from $1 \leq n \leq 16384$

S = size of sample

T = number of integers in N, satisfying conditions of theorem 3

E = rounded elimination percentage

N	S	T	E
All n	16384	4106	75%
$n \equiv 0 \bmod 4$	4096	264	94%
$n \equiv 1 \bmod 4$	4096	1034	75%
$n \equiv 2 \bmod 4$	4096	187	95%
$n \equiv 3 \bmod 4$	4096	2621	36%
n even and $\alpha(n) \equiv 0 \bmod 4$	2015	178	91%

N	S	T	E
$\alpha(n) = 4$	1001	56	94%
$\alpha(n) = 5$	2002	550	73%
$\alpha(n) = 6$	3003	537	82%
$\alpha(n) = 7$	3432	740	78%
$\alpha(n) = 8$	3003	399	87%
$\alpha(n) = 9$	2002	1060	47%
$\alpha(n) = 10$	1001	462	54%
$\alpha(n) = 11$	364	216	41%
$\alpha(n) = 12$	91	54	41%

§5. Explicit computations for $\alpha(n) \leq 5$

We list below, with some appropriate comments, the explicit conditions imposed on n by theorem 3.

A. $\alpha(n) = 2$

$$(\log(1+t)/t)^{2n-1} = 1 - \frac{2n-1}{2} t + \frac{(2n-1)(3n+1)}{12} t^2 - \cdots.$$

Conditions:

$$a_o:\ \nu_2(e_o) \geq 1$$

$$a_1:\ \nu_2(e_o) - 1 = 0$$

$$a_2:\ \nu_2(e_o) + \nu_2(3n+1) - 2 = 0$$

Solution: $n \equiv 3 \pmod 4$, $\nu_2(e_o) = 1$. As $\alpha(n) = 2$, the only possible value of n is $n = 3$.

The immersion $CP^3 \looparrowright R^8$ is nevertheless impossible. We leave this as an exercise. (Hint: use $\psi^3(\mu_o) \equiv \mu_o^3 \pmod 3$).

B. $\alpha(n) = 3$

$$(\log(1+t)/t)^{2n-2} = 1-(n-1)t+\frac{(n-1)(6n-1)}{12} t^2 - \frac{n(n-1)(2n+1)}{12} t^3 + \cdots$$

Conditions:

a_o: $\nu_2(e_o)\geq 1$

a_1: $\nu_2(e_o)+\nu_2(n-1)\geq 1$

a_2: $\nu_2(e_o)+\nu_2(n-1)-2=0$

a_3: $\nu_2(e_o)+\nu_2(n)+\nu_2(n-1)-2=0$

Solutions: $n \equiv 3 \pmod 4$, $\nu_2(e_o) = 1$. As $\alpha(n) = 3$ the only possible values of n are $n = 2^s+3$, $s\geq 2$.

Precisely in these cases, an immersion is given by Steer (theorem D).

C. $\alpha(n) = 4$

$$(\log(1+t)/t)^{2n-3} = 1 - \frac{2n-3}{2}t + \frac{(2n-3)(3n-2)}{12}t^2 - \frac{n(2n-3)(2n-1)}{24}t^3 + \frac{(2n-3)(30n^3+15n^2-5n-2)}{1440}t^4 - \dots$$

Conditions:

a_o: $\nu_2(e_o)\geq 1$

a_1: $\nu_2(e_o)-1\geq 1$

a_2: $\nu_2(e_o)+\nu_2(3n-2)-2\geq 1$

a_3: $\nu_2(e_o)+\nu_2(n)-3=0$

a_4: $\nu_2(e_o)+\nu_2(30n^3+15n^2-5n-2)-5=0$

Solutions: 1) $n \equiv 7 \bmod 8$, $\nu_2(e_o)=3$ ($n=2^s+7$, $s\geq 3$)

2) $n \equiv 10 \bmod 16$, $\nu_2(e_o)=2$

D. $\alpha(n) = 5$

We content ourselves with the result, as the computations are entirely analogous. The solutions are:

1) $n \equiv 1 \bmod 4$, $\nu_2(e_o) = 4$

2) $n \equiv 7 \bmod 8$, $\nu_2(e_o) = 3$

§6. Some general results

We now give a consequence of theorem 3 which produces a directly computable condition on n. As we could check on our sample, this condition is weaker, but brings into play the function $\alpha(\alpha(n))$. For this reason we think it may be a useful information.

Theorem 4: If $CP^n \looparrowright R^{4n-2\alpha(n)}$, then $\binom{n+1}{\alpha(n)} \equiv 0 \pmod 2$

Proof: The equation $a_{\alpha(n)-1} \equiv a_{\alpha(n)} \equiv 1 \pmod 2$ reads

$$\frac{e_o\cdot(2n-\alpha(n)+1)!}{(2n+1)!} S(2n+1,2n-\alpha(n)+1) \equiv \frac{e_o\cdot(2n-\alpha(n)+1)!}{(2n)!} S(2n,2n-\alpha(n)+1) \equiv 1 \pmod 2.$$

This implies in particular:

$$S(2n+1,2n-\alpha(n)+1) \equiv S(2n,2n-\alpha(n)+1) \pmod 2$$

To study this congruence, we use an easy combinatorial result, whose routine proof is left to the reader:

Lemma: $S(m+q,m) \equiv \binom{m+q}{2q} \pmod 2$

We therefore deduce successively

$$\binom{2n+1}{2\alpha(n)} \equiv \binom{2n}{2\alpha(n)-2} \pmod 2 \Rightarrow \binom{n}{\alpha(n)} \equiv \binom{n}{\alpha(n)-1} \pmod 2 \Rightarrow \binom{n+1}{\alpha(n)} \equiv 0 \pmod 2, \text{ q.e.d.}$$

Computing with dyadic expansions, it is now very easy to produce infinite families of non-immersion values of n.

Corollary: $CP^n \not\hookrightarrow R^{4n-2\alpha(n)}$ for the following values of n:

$2^s-2,\ 2^{2s+1}-3,\ 2^{4s+2}-4,\ 2^{4s+3}-4,\ 2^{4s+1}-5,\ 2^{8s}-6,\ 2^{8s+1}-6,\ 2^{8s+2}-6,\ 2^{8s+3}-6.$

To formulate our next result with complete precision would require the explicit form of the James numbers, as well as some combinatorial manipulations. We shall avoid this in being intentionally imprecise, because we feel the type of results is in itself the interesting phenomenon.

Theorem 5: The immersion $CP^n \hookrightarrow R^{4n-2\alpha(n)}$ is impossible in the following two situations:

(i) $2n-\alpha(n)+1$ is divisible by a high power of 2.

(ii) n is divisible by a high power of 2.

Proof:

(i) The necessary power of 2 is the one guaranteeing that the relevant coefficients of $(\log(1+t)/t)^{2n-\alpha(n)+1}$ have odd denominators. Then, the integer e_o in theorem 3 cannot be even, proving the result.

(ii) Similarly, we make sure that the corresponding coefficients of the two series $(\log(1+t)/t)^{2n-\alpha(n)+1}$ and $(\log(1+t)/t)^{-\alpha(n)+1}$ are divisible by the same power of 2. Now, for the second series, the residue theorem gives easily:

$$\text{-coefficient of } t^{\alpha(n)-1} = \pm \text{ coefficient of } z^{\alpha(n)-1} \text{ in } \frac{z}{e^z-1}$$

$$- \text{coefficient of } t^{\alpha(n)} = \text{coefficient of } z^{\alpha(n)} \text{ in } \left(\frac{z}{2} / \text{sh}\frac{z}{2}\right)^2$$

A trivial parity argument shows then that the condition $a_{\alpha(n)-1} \equiv a_{\alpha(n)} \equiv 1 \pmod 2$ of theorem 3 is impossible to fulfill. This completes the proof.

References

1. J.F. Adams and G. Walker, On complex Stiefel manifolds, Proc. Cambridge Philos. Soc. 61(1965)81-103.

2. M.F. Atiyah, Thom complexes, Proc. London Math. Soc. (3)11(1961)291-310.

3. M. F. Atiyah et F. Hirzebruch, Quelques théorèmes de non-plongement pour les variétés différentiables, Bull. Soc. Math. France 87(1959)383-396.

4. M.F. Atiyah and J.A. Todd, On complex Stiefel manifolds, Proc. Cambridge Philos. Soc. 56(1960)342-353.

5. I.M. James, Euclidean models of projective spaces, Bull. London Math. Soc. 3(1971)257-276.

6. K.H. Mayer, Elliptische Differentialoperatoren und Ganzzahligkeitssätze für charakteristische Zahlen, Topology 4(1965)295-313.

7. B.J. Sanderson and R.L.E. Schwarzenberger, Non-immersion theorems for differentiable manifolds, Proc. Cambridge Philos. Soc. 59(1963)312-322.

8. B. Steer, On the embedding of projective spaces in Euclidean space, Proc. London Math. Soc. (3)21(1970)489-501.

9. B. Steer, On immersing complex projective (4k+3)-space in Euclidean space, Quart. J. Math. Oxford (2)22(1971)339-345.

ON THE EXPONENT AND THE ORDER OF THE GROUPS $\tilde{J}(X)$

By

François Sigrist and Ueli Suter

1. INTRODUCTION

Let $\tilde{J}(X)$ be the group of equivalence classes of orthogonal sphere bundles over a CW-complex X with respect to stable fibre homotopy type. It is well known that $\tilde{J}(X)$ is finite if X is a finite CW-complex [4]. On the other hand, the solution of the complex vector field problem on spheres by Adams and Walker [3] provides elements of surprisingly high order in the group $\tilde{J}(CP^m)$. Explicitely, Adams and Walker show that the $\tilde{J}$-order of the Hopf bundle ξ over CP^m equals the so called complex James number $b(m+1)$, a number divisible by $(m+1)!$ and roughly of the same order of magnitude [8].

This raises the question of existence of high order elements in $\tilde{J}(X)$, and more generally, the question of the "size" of the groups $\tilde{J}(X)$. The purpose of this paper is to give upper bounds for the exponent and the order of these groups.

The following two theorems, which result directly from our stronger but more technical theorems 1 and 2 (see §2 and §3 respectively), illustrate the type of bounds we will obtain.

Theorem 1': Let X be a CW-complex of finite dimension $n \geq 4$. Then, the exponent of the group $\tilde{J}(X)$ divides the complex James number $b([\frac{1}{2}n]+1)$. In particular we have $b([\frac{1}{2}n]+1)\cdot\tilde{J}(X)=0$.

Theorem 2': Let X be a 2-torsion free finite CW-complex. Then, the order of the group $\tilde{J}(X)$ divides the product

$$\prod_{q=1}^{\dim(X)} |H^q(X;Z)\otimes\tilde{J}(S^q)|.$$

Obvious examples show that both bounds cannot be improved in general.

To prove theorem 1 (see §2), we first reduce the general problem to the computation of the $\tilde{J}$-order of O(2)-bundles, using a slight refinement of the (already refined) splitting principle of Becker and Gottlieb [5], and then conclude by representation theory. Theorem 2 (see §3) is proved by induction on the number of cyclic summands in a direct sum decomposition of $H_*(X;Z)$, starting with the $\tilde{J}$-groups of Moore spaces $K'(C,q)$ with C cyclic.

Throughout this paper we make use of the Adams Conjecture first proved by Quillen in 1970 [7], i.e. we will identify the group $\tilde{J}(X)$ with the group $\tilde{J}''(X)$ introduced by Adams [2].

In order to avoid unnecessary confusions, we shall suppose in this paper that CW-spaces are connected; the general case requires only minor and well-known adjustments.

2. THE EXPONENT OF $\tilde{J}(X)$

We will give an upper bound for the exponent of $\tilde{J}(X)$ in terms of the real and complex James numbers we respectively denote by $a(n+1)$ and $b(m+1)$. Those are determined as follows (see [1] and [3]):

$$a(n+1) = |\widetilde{KO}(RP^n)| = |\tilde{J}(RP^n)| = 2^{f(n)}, \text{ with } f(n) \text{ given by}$$

$f(n+8) = f(n)+4$ and

n	1	2	3	4	5	6	7	8
f(n)	1	2	2	3	3	3	3	4

$$\nu_p(b(m+1)) = \begin{cases} \max\limits_r (r+\nu_p(r)),\ 1 \leq r \leq \left[\frac{m}{p-1}\right] & p \leq m+1 \\ 0 & p > m+1 \end{cases}$$

($\nu_p(c)$ denotes the exponent of the prime p in the prime power decomposition of c).

Theorem 1: Let X be an n-dimensional CW-complex. Then the exponent of the group $\tilde{J}(X)$ divides the l.c.m. of the integers $a(n+1)$ and $b([\frac{1}{2}n]+1)$, which is $a(n+1)$ for $n \leq 3$ and $b([\frac{1}{2}n]+1)$ for $n \geq 4$.

Before giving the proof, we provide two Lemmas.

Lemma 1: Let X be a finite CW-complex of dimension n and ξ an $O(2q)$-bundle over X. Then, there exist firstly

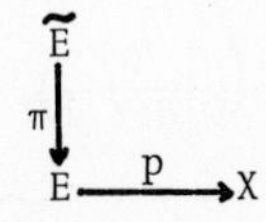

with E an n-dimensional CW-complex and π a finite covering map, and secondly an $O(2)$-bundle ζ over $\tilde{E}$, such that

(i) $p^* : \tilde{J}(X) \to \tilde{J}(E)$ is a monomorphism

(ii) $p^*(\xi) = \pi_*(\zeta)$, where π_* is the transfer associated to the covering π.

Remark: Without the dimension condition on E, this Lemma is the refined splitting principle of Becker and Gottlieb [5, §7].

Proof: By Becker-Gottlieb (op.cit.) we get $\tilde{E}'$, E', p', π', ζ' satisfying conditions (i) and (ii) in the Lemma. Taking now $E = (E')^n$, the n-skeleton of E', we look at

$$\begin{array}{ccccc} \tilde{E} & \xrightarrow{j} & \tilde{E}' & & \\ \downarrow{\scriptstyle\pi} & & \downarrow{\scriptstyle\pi'} & & \\ E & \xrightarrow{i} & E' & \xrightarrow{p'} & X \end{array}$$

where π is the induced covering. We set $p=p'\cdot i$ and $\zeta=j^*(\zeta')$. By naturality of the covering transfer, we immediately get $p^*(\xi) = \pi_*(\zeta)$. It remains to show that p induces a monomorphism of the $\tilde{J}$-groups. Again by [5], p' is a fibration having a stable section $\tau' : \Sigma^q(X^+) \to \Sigma^q(E'^+)$. By cellular approximation, τ' factors homotopically through a map $\tau: \Sigma^q(X^+) \to \Sigma^q(E^+)$, which provides a stable section of $p: E \to X$. As a stable map induces a natural homomorphism at the $\tilde{J}$"-level, the conclusion follows.

Lemma 2: Let ζ be an O(2)-bundle over a finite CW-complex X of dimension n. Then the order of $\tilde{J}(\zeta)$ divides the l.c.m. of $a(n+1)$ and $b([\tfrac{1}{2}n]+1)$.

Proof: The bundle ζ is classified by a map from X into the n-skeleton B^n of the classifying space BO(2). Hence it suffices to prove the Lemma for the n-universal bundle ζ_n over B^n.

We make use of some well-known relations between KO-theory and representation theory. The determinant $O(2) \to Z/2$ gives rise to a map $BO(2) \to BZ/2 = RP^\infty$ and hence to a map $d: B^n \to RP^n$ inducing $d^!: KO(RP^n) \to KO(B^n)$. Let β_n be the canonical line bundle over RP^n and $\delta_n = d^!(\beta_n)$ its image in $KO(B^n)$. The representation $O(2) \to SO(3)$ given by

$$A \to \left(\begin{array}{c|c} A & 0 \\ \hline 0 & \det a \end{array}\right)$$

induces a map $g: B^n \to BSO(3)$, and together with the α-construction [6,12(5.4)] we get the ψ-ring homomorphisms

$$\begin{array}{ccc} RO(SO(3)) & \xrightarrow{h = g^! \cdot \alpha} & KO(B^n) \\ & {\scriptstyle\alpha}\searrow \quad \nearrow{\scriptstyle g^!} & \\ & KO(BSO(3)) & \end{array}$$

$RO(SO(3))$ being the real representation ring of the Lie group $SO(3)$. For the canonical representation λ in $RO(SO(3))$ we have

$$h(\lambda) = \zeta_n + \delta_n \text{ in } KO(B^n).$$

Hence the $\tilde{J}$-order of ζ_n divides the l.c.m. of $|\tilde{J}(h(\lambda))|$ and $|\tilde{J}(\delta_n)|$. The latter divides $|\tilde{J}(B_n)| = a(n+1)$, and Lemma 2 will be proved if we can show that $|\tilde{J}(h(\lambda))|$ divides $b([\frac{1}{2}n]+1)$.

We set $\lambda = 3 + \omega$ in $RO(SO(3))$, and note [6,13(12.1)] $RO(SO(3)) \cong Z[\omega]$. The Adams operations on $RO(SO(3))$ are given by $\psi^k(\omega)=T_k(\omega)$, where T_k is the polynomial introduced by Adams and Walker in [3]. Investigating the filtration on various elements in the filtered ring $KO(B^n)$, we show now that h factors through a ψ-quotient Q_n of $Z[\omega]$, which is isomorphic to $KO(CP^{[\frac{1}{2}n]})$. From the Atiyah-Hirzebruch spectral sequence of $BSO(3)$ we infer

$$\text{filt}(\alpha(\omega)) \geq 2, \ \text{filt}(2\alpha(\omega)) \geq 4, \ \text{filt}\ (\alpha(\omega)^2) \geq 8.$$

The homomorphism $g^!$ is compatible with filtration and we get for the element $h(\omega) = g^! \cdot \alpha(\omega)$:

$$\text{filt}(h(\omega)^{2q}) \geq 8q, \ \text{filt}(h(\omega)^{2q+1}) \geq 8q+2, \ \text{filt}(2h(\omega)^{2q+1}) \geq 8q+4.$$

Since in $KO(B^n)$ elements of filtration $>n$ are zero, we conclude that h induces a ψ-ring homomorphism $\bar{h}: Q_n \longrightarrow KO(B^n)$, where

$$Q_n = \begin{cases} Z[\omega]/(\omega^{m+1}), & m = [\frac{1}{2}n] \not\equiv 1 \pmod 4 \\ Z[\omega]/(\omega^{m+1}, 2\omega^m), & m = [\frac{1}{2}n] \equiv 1 \pmod 4. \end{cases}$$

Consulting Adams-Walker [3] we see that the order of the element $\tilde{J}''(\bar{\omega})$ in $J''(Q_n)$ is equal to $b([\frac{1}{2}n]+1)$. The homomorphism $\tilde{J}''(\bar{h}): \tilde{J}''(Q_n) \to \tilde{J}''(B^n) = \tilde{J}(B^n)$ sends $\tilde{J}''(\bar{\omega})$ onto $\tilde{J}''(h(\omega)) = \tilde{J}''(h(\lambda))$ whose order therefore divides $b([\frac{1}{2}n]+1)$. Lemma 2 is now proved.

<u>Proof of theorem 1:</u> Let X be an n-dimensional CW-complex and let $\mu \varepsilon \tilde{J}(X)$. The element μ can be represented by an $O(2q)$-bundle ξ, $2q>n+1$. We set $c_n = \text{l.c.m.}(a(n+1), b([\frac{1}{2}n]+1))$. If X is a finite complex, applying Lemma 1

and noting that the transfer π_*: $KO(\tilde{E}) \to KO(E)$ induces a homomorphism at the $\tilde{J}''$-level, we conclude with Lemma 2 that $c_n \cdot \tilde{J}(\xi) = 0$. Suppose now that X is an arbitrary CW-complex of dimension n. The bundle ξ is induced from the n-universal $O(2q)$-bundle ω_n by a map $g: X \to (BO(2q))^n$ (n-skeleton). The latter space is a finite complex and we get $c_n \cdot \tilde{J}(\xi) = g^*(c_n \cdot \tilde{J}(\omega_n)) = 0$. This completes the proof.

3. THE ORDER OF $\tilde{J}(X)$

We first list the $\tilde{J}$-groups of the Moore spaces $K'(C,q)$ with C cyclic. The space $K'(Z,q)$ is the sphere S^q and we have, according to Adams [2, §3]:

$$\tilde{J}(S^q) = \begin{cases} Z/m(2t) & q = 4t > 0 \\ Z/2 & q \equiv 1,2 \pmod 8 \\ 0 & \text{otherwise} \end{cases}$$

The integer $m(2t)$ is the g.c.d. of the numbers $k^N(k^{2t}-1)$, $k \in \mathbb{N}$, N large. It is explicitely determined in [2, §2].

In the following Lemma we write $\widetilde{KO}(Z/k,q)$ for the reduced KO-theory and $\tilde{J}(Z/k,q)$ for the $\tilde{J}$-group of the space $K'(Z/k,q)$.

<u>Lemma:</u>

$$\tilde{J}(Z/k,q) = 0 \qquad q \equiv 4,5,6 \pmod 8$$

$$\tilde{J}(Z/k,4t-1) = Z/(m(2t),k) = Z/k \otimes \tilde{J}(S^{4t}) \qquad (t > 0)$$

$$\tilde{J}(Z/k,8t) = \widetilde{KO}(Z/k,8t) = Z/k \otimes Z/2 \qquad (t > 0)$$

$$\tilde{J}(Z/k,8t+2) = \widetilde{KO}(Z/k,8t+2) = Z/k \otimes Z/2 \qquad (t \geq 0)$$

$$\tilde{J}(Z/k,8t+1) = \widetilde{KO}(Z/k,8t+1) = \begin{cases} Z/4 & k \equiv 2 \pmod 4 \\ Z/2 \oplus Z/2 & k \equiv 0 \pmod 4 \\ 0 & k \text{ odd } (t > 0) \end{cases}$$

<u>Proof:</u> The space $K'(Z/k,q)$ can be given as the mapping cone of a degree k map from S^q to S^q. The KO-theory Puppe sequence easily yields the ψ-group $\widetilde{KO}(Z/k,q)$ from the ψ-groups $\widetilde{KO}(S^{q+1})$ and $\widetilde{KO}(S^q)$ which are well-known [2]. The details are almost straightforward, and can be safely left out.

We will give an upper bound of the order $|\tilde{J}(X)|$ in terms of the integral homology $H_*(X)$ of X. Let $H_q(X) = F_q(X) \oplus T_q(X)$ where $F_q(X)$ is a free part and $T_q(X)$ the torsion part of the q-th homology group. Call f_q the rank of $F_q(X)$.

Theorem 2: Let X be a finite complex of dimension n and let $k_q = |\tilde{J}(S^q)|^{f_q} \cdot t_q$, with

$$t_q = \begin{cases} 1 & q \equiv 4,5,6 \pmod 8 \\ |T_q(X) \otimes \tilde{J}(S^{q+1})| & q \equiv 0,3,7 \pmod 8 \\ |T_q(X) \otimes Z/2| & q \equiv 2 \pmod 8 \\ |T_q(X) \otimes Z/2|^2 & q \equiv 1 \pmod 8 \end{cases}$$

Then, $|\tilde{J}(X)|$ divides $k(X) = k_1 \cdot k_2 \cdots k_n$

Proof: (a) We first prove the Theorem for simply-connected CW-complexes, by induction on the number of cyclic summands in a direct sum decomposition of the abelian group $\tilde{H}_*(X)$. If $\tilde{H}_*(X)$ has only one cyclic summand, X is a Moore space $K'(C,q)$ with C cyclic. From the above Lemma we infer $|\tilde{J}(K'(C,q))| = k(K'(C,q))$. Suppose then that $H_*(X)$ is decomposed into m+1 direct cyclic summands. In the first non-zero homology group $H_q(X)$ we choose a direct cyclic summand C. Since X is simply-connected we have $q \geq 2$, and there is a map $f: K'(C,q) \to X$ such that $\operatorname{Im} f_* = C$. The cofibre sequence $K'(C,q) \xrightarrow{f} X \xrightarrow{g} C_f$ gives rise to a decomposition $\tilde{H}_*(X) = \tilde{H}_*(K'(C,q)) \oplus \tilde{H}_*(C_f)$ and hence $k(X) = k(C_f) \cdot k(K'(C,q))$. From the exact sequence of finitely generated ψ-groups

$$\widetilde{KO}(C_f) \xrightarrow{g^!} \widetilde{KO}(X) \to \operatorname{Im} f^! \to 0$$

we get by [2,3.8] the exact sequence

$$\tilde{J}''(C_f) \to \tilde{J}''(X) \to \tilde{J}''(\operatorname{Im} f^!) \to 0$$

Thus $|\tilde{J}''(X)|$ divides $|\tilde{J}''(C_f)| \cdot |\tilde{J}''(\operatorname{Im} f^!)|$. $\tilde{H}_*(C_f)$ having m summands, we know by inductive hypothesis that $|\tilde{J}(C_f)|$ is a divisor of $k(C_f)$. Case by case inspection shows then that $|\tilde{J}''(\operatorname{Im} f^!)|$ divides $k(K'(C,q))$. This proves the Theorem for simply-connected CW-complexes.

(b) Let now X be any finite CW-complex of dimension n. There is a simply-connected CW-complex X_1 and a map $h: X \to X_1$ inducing homology isomorphisms in dimensions ≥ 2. We infer $k(X) = k_1 \cdot k(X_1)$. Using again the right-exactness of $\tilde{J}''$, we look at

$$\tilde{J}''(X_1) \to \tilde{J}''(X) \to \tilde{J}''(\operatorname{Coker} h^!) \to 0$$

It is easily seen that $|\operatorname{Coker} h^!|$ divides k_1. So does a fortiori $|\tilde{J}''(\operatorname{Coker} h^!)|$. As by (a) the order of $\tilde{J}''(X_1)$ divides $k(X_1)$, the proof of Theorem 2 is complete.

References

1. J.F. Adams, Vector fields on spheres, Ann. of Math. 75(1962)603-632.

2. J.F. Adams, On the groups J(X)-II, Topology 3(1965)137-171.

3. J.F. Adams and G. Walker, On complex Stiefel manifolds, Proc. Cambridge Philos. Soc. 61(1965)81-103.

4. M.F. Atiyah, Thom complexes, Proc. London Math. Soc. (3)11(1961)291-310.

5. J.C. Becker and D.H. Gottlieb, The transfer map and fiber bundles, Topology 14(1975)1-12.

6. D. Husemoller, Fibre bundles, McGraw-Hill Book Company, New-York (1966).

7. D. Quillen, The Adams Conjecture, Topology 10(1970)67-80.

8. F. Sigrist, Deux propriétés des groupes $J(CP^{2n})$, Rendic. Acc. Naz. Lincei (8)49(1975)413-415.

STABLE DECOMPOSITIONS OF CLASSIFYING SPACES WITH APPLICATIONS TO ALGEBRAIC COBORDISM THEORIES

VICTOR SNAITH†

§1. Introduction

This paper will describe how one may decompose the S-type of classifying spaces such as BU, BO, BSp, $BO\,F_q$ and $BGL\,F_q$. The decompositions are all very similar. For example, there is an S-equivalence

$$BU \simeq \bigvee_{1 \leq k} MU(k). \tag{1.1}$$

The decompositions will be presented in detail and proved in §2. I am told that the existence of the S-equivalence (1.1) was formerly a conjecture of A. Liulevicius.

Let $\{A,B\}$ denote S-homotopy classes of S-maps from A to B[1]. It is well-known that if k is sufficiently large in comparison with dim X then $\{X,MU(k)\}$ is isomorphic to $MU^{2k}(X)$, the 2k-th unitary cobordism group of X. This observation suggests that, by suitably refining (1.1), we may obtain $MU^{2*}(X)$, through a range of dimensions, in terms of S-maps from X to BU. This, and its real and symplectic analogues, are formalised and proved in §3. The first application between $\pi_*^S(BU)$ and $\pi_*(MU)$ is to compute a few of the groups

$$\pi_j^S(BU)/(\text{odd torsion}).$$

This computation is explained in §4, which also contains some problems suggested by and/or motivating the material of §§2-3.

It has been known for some time that unitary cobordism determines KU-theory [8]. In fact a new proof of this is given in §8. This means that the S-types MU(k) $(1 \leq k)$ determine BU, the classifying space, in a canonical manner. However the results of §3 suggest -- rather remarkably -- the "converse", that BU determines MU-theory. In §5 this "converse" is proved, together with its real and symplectic

† Research partially supported by a grant from the Canadian N.R.C.

analogues. It is shown, for example, how a simple construction in terms of BU yields a ring, AU(X), which is naturally isomorphic to $MU^{2*}(X)$ when $\dim X < \infty$. In §6 this construction is formalised to give a spectrum in return for a homotopy commutative, homotopy associative H-space and a homogeneous element of its stable homotopy. To BU and a generator of $\pi_2(BU)$ is associated the AU-spectrum mentioned above. Also in §6 a miscellany of further examples are given. These include the AO- and AS_p-spectrum, which give MO*- and MSp^{4*}-theory, and the algebraic cobordism spectra which may be associated with a ring.

In §7 we take a look at AZ-theory, the algebraic cobordism theory of Z (the integers). Being disconnected the algebraic cobordism theories are very difficult to calculate. However, in §7 it will be shown that there is a natural homomorphism

$$AX^o(X) \to MO^*(X)$$

which is onto if $\dim X < \infty$ and is not an isomorphism when $X = S^n$, the n-sphere.

In §8 a closer look is taken at the AU-spectrum in terms of which several "classical" aspects of MU-theory are described. A particularly nice point occurs here when one finds that Adams' idempotent in MU-theory [2] is constructible without any knowledge of $\pi_*(MU)$. The other MU-theory phenomena treated in §8 will be Adams operations, Landweber-Novikov operations, the complexification homomorphism $c: MSp^*(\_) \to MU^*(\_)$, the Thom isomorphism, the Conner-Floyd isomorphism, the Hattori-Stong theorem and the Pontrjagin-Thom construction.

In §9 the canonical étale algebraic cobordism theory associated with k, an algebraically closed field of finite characteristic, is identified as $M\hat{UZ}^{2*}$-theory. It is shown how to imitate the Pontrjagin-Thom construction for a smooth algebraic embedding over k. To the cognoscenti §9 will be seen to yield no more invariants of the smooth algebraic embedding than one may obtain by applying $M\hat{UZ}^*$-theory to the étale homotopy type of the embedding. However, the point of §9 is to show how the process may be carried out entirely in terms of the transfer construction. Hence one may (perhaps optimistically) hope to form similar Pontrjagin-Thom constructions for an arbitrary smooth algebraic embedding. The major fault to be found with the approach of §9 is that there is too much homotopy theory and too little by way of direct geometrical construction.

This paper is an elaboration upon two lectures given in Vancouver at the Canadian Mathematical Congress meeting held in July and August 1977. I am very grateful to the organisers for this opportunity to describe the material. I hope that the main body of this work will appear in full detail in [20; 21 and 22]. I have therefore given only sketches of proofs. In the body of the text I have included several problems, each to be found in the vicinity of the material to which it refers.

§2. Decomposition of some S-types

2.1 We will require two basic pieces of apparatus -- some stable decomposition of $\Omega^n\Sigma^n X$ and the transfer construction.

Firstly we review the decompositions. Details are to be found in [23]. However C. Reedy [19] subsequently gave a semi-simplical proof and recently F. Cohen and L. Taylor [7] have given a simpler version of the original decomposition maps in terms of configuration spaces.

When X is a connected, pointed CW complex with a compactly generated topology there exists a nice model, denoted by C_nX, for the space $\Omega^n\Sigma^n X$ up to homotopy ($1 \le n \le \infty$). C_nX is a filtered space

$$C_nX \supset \cdots \supset F_mC_nX \supset F_{m-1}C_nX \supset \cdots \supset F_1C_nX = X. \tag{2.2}$$

The filtration (2.2) is compatible with the suspension map $\Omega^n\Sigma^n X \to \Omega^{n+1}\Sigma^{n+1}X$. For each $1 \le m,n \le \infty$ there are S-equivalences

$$F_mC_nX \simeq \bigvee_{1 \le k \le m} \frac{F_kC_nX}{F_{k-1}C_nX} . \tag{2.3}$$

The S-equivalences of (2.3) are compatible as m and n vary. Also the S-map induced by coming off onto $\frac{F_mC_nX}{F_{m-1}C_nX}$ in (2.3) is S-homotopic to the canonical collapsing map. We will need to know that when $n = \infty$ and $X = BH$, the classifying space of a compact group H, then $F_kC_\infty BH$ is a quotient of

$$E\Sigma_k \underset{\Sigma_k}{\times} (BH)^k \cong B\Sigma_k \int H$$

and that there are homeomorphisms

(2.4)
$$\frac{F_k C_\infty BH}{F_{k-1} C_\infty BH} \cong \frac{B\Sigma_k \int H}{B\Sigma_{k-1} \int H} .$$

Here $\Sigma_k \int H$ is the wreath product generated by the normal subgroup H^k and the symmetric group, Σ_k, which acts on H^k by permuting the factors.

Now let us turn to the transfer for which the basic references are [3;4;15]. For a fibration $F \to E \xrightarrow{\pi} B$ in which F has the homotopy type of a compact space the transfer is an element

(2.5)
$$\tau(\pi) \in \{B,E\} \cong [B,QE].$$

Here $[\_,\_]$ denotes based homotopy classes and

$$QE = \lim_{\overrightarrow{n}} \Omega^n \Sigma^n E.$$

$\tau(\pi)$ is natural for induced fibrations and $\pi_* \circ \tau(\pi)_*: H_*(B) \to H_*(E) \to H_*(B)$ is multiplication by the Euler characteristic of F. Also we have the Double Coset Formula [10]. Let $j: H \subset G$ and $k: K \subset G$ be inclusions of compact Lie groups. For $g \in G$ let $\sigma_g \in \{BK, BH\}$ denote the composite

(2.6)
$$BK \xrightarrow{\text{transfer}} B(K \cap gHg^{-1}) \to B(gHg^{-1}) \to BH.$$

In (2.6) the second map is induced by group inclusion and the third by conjugation by g. Consider the following diagram.

$$\begin{array}{ccccc} & & & & BK \\ & & & & \downarrow BK \\ G/H & \to & BH & \xrightarrow{B_j} & BG \end{array}$$

The Double Coset Formula asserts

(2.7)
$$\tau(B_j) \circ B_k = \Sigma\, \sigma_g$$

where the sum is taken over double coset representatives for $K \backslash G/H$ and equality is asserted only when the right hand side makes sense.

In [6] and [20, Part I] formulae generalising (2.7) are obtained using differential geometry -- in the case of smooth fibre bundles.

.8 Remark. Before proceeding to describe my decomposition maps I should like to ecord some other recent applications of the stable decompositions of (2.3). Using he case $n = 2$ and $X = S^7$ Mahowald has constructed the much sought after elements f $\pi_j^S(S^o)$ represented by h_1h_j $(j \neq 2)$ in the classical mod 2 Adams spectral equence. Related to his work are stable decompositions of the factors

$$\frac{F_k C_2 S^m}{F_{k-1} C_2 S^m}$$

nto Brown-Gitler spectra. These still more recent results are due to E.H. Brown Jr. nd F. Peterson. In fact Mahowald's original construction of h_1h_j contained a gap t this point, namely the assumption that

$$\frac{F_k C_2 S^{2t+1}}{F_{k-1} C_2 S^{2t+1}}$$

s a Brown-Gitler spectrum.

.9 The S-decomposition maps. We may form the following composition of maps and S-maps

$$BU(n) \xrightarrow{\tau(\pi_n)} B\Sigma_n \int U(1) \to F_n C_\infty BU(1) \simeq \bigvee_{1\le k\le n} \frac{F_k C_\infty BU(1)}{F_{k-1} C_\infty BU(1)}$$

(2.10)

$$\to \bigvee_{1\le k\le n} \frac{B\Sigma_k \int U(1)}{B\Sigma_{k-1}\int U(1)} \longrightarrow \bigvee_{1\le k\le n} \frac{BU(k)}{BU(k-1)} .$$

Denote by $\nu_{U(n)}$ the S-map of (2.10). Here $\tau(\pi_n)$ is the transfer associated with the fibration

$$U(n)/\Sigma_n \int U(1) \to B\Sigma_n \int U(1) \xrightarrow{\pi_n} BU(1),$$

the second map is the quotient map mentioned above, the third is (2.3), the fourth is (2.4) and the last map is induced by the maps $(\pi_k;\ 1 \le k \le n)$.

There are similar S-map compositions.

$$\nu_{Sp(n)}: BSp(n) \longrightarrow \bigvee_{1\le k\le n} \frac{BSp(k)}{BSp(k-1)}$$

(2.11)

$$\nu_{O(2n)}: BO(2n) \longrightarrow \bigvee_{1\le k\le n} \frac{BO(2k)}{BO(2k-2)} \quad .$$

These are obtained by replacing $(U(1),U(n))$ by $(Sp(1),Sp(n))$ or by $(O(2),O(2n))$ as appropriate.

<u>2.12 Theorem</u>. If $G_n = U(n)$, $Sp(n)$ or $O(2n)$ $(1 \le n \le \infty)$ then

(i) ν_{G_n} is an S-equivalence

and

(ii) $\nu_{G_n}|BG_{n-1} \simeq \nu_{G_{n-1}}$ (i.e. $\{\nu_{G_n}\}$ is compatible as n varies).

<u>Proof</u>: The compatibility statement (ii) reduces to showing, for example, that

$$\begin{array}{ccc} BU(n-1) & \xrightarrow{\tau(\pi_{n-1})} & B\Sigma_{n-1}\int U(1) \\ \downarrow & & \downarrow \\ BU(n) & \xrightarrow[\tau(\pi_n)]{} & B\Sigma_n\int U(1) \end{array}$$

is S-homotopy commutative. The vertical maps are induced by the canonical group inclusions. However in this case the Double Coset Formula (2.7) has just one term, which expresses the required homotopy commutativity.

For (i) one has to check that ν_{G_n} induces a homology isomorphism. However this may be accomplished easily by observing that $G_n/\Sigma_n\int G_1$ has Euler characteristic equal to one and appealing to the properties of the decomposition (2.3) which were listed above.

<u>2.13</u>. Similar stable decomposition of $BO\,F_3$ and $BGL\,F_q$ will be given. Let F_q be the field with q elements. Let ℓ be a prime not dividing q and set r equal to the order of q in the units $(Z/\ell)^*$. Let $GL_n F_q$ be the general linear group of $n \times n$ matrices over F_q. Also let $O_n F_3$ be the subgroup of $GL_n F_3$ which preserves the form $\sum_{i=1}^{n} \alpha_i^2$.

2.14. Let q, ℓ, r be as in §2.13. Then for $1 \le n \le \infty$ there is an ℓ-local S-equivalence

$$BGL_{nM} F_q \simeq \bigvee_{1 \le k \le n} \frac{BGL_{kM} F_q}{BGL_{(k-1)M} F_q}$$

where

$$M = \begin{cases} r & \text{if } \ell \ne 2 \text{ or } \ell = 2,\ q \equiv 1(4) \\ 2 & \text{otherwise} \end{cases}.$$

2.15 Theorem. For $1 \le n \le \infty$ there exists a 2-local S-equivalence

$$\phi_n : BO_{2n} F_3 \xrightarrow{\simeq} \bigvee_{1 \le k \le n} \frac{BO_{2k} F_3}{BO_{2k-2} F_3}.$$

Proof of 2.15: (The proof of 2.14 is similar.) There is just one difficulty in replacing $(U(1), U(n))$ by $(O_2 F_3, O_{2n} F_3)$ in the proof of 2.12. We do not know that the transfers associated with the maps

$$B\pi_n : B\Sigma_n \int O_2 F_3 \longrightarrow BO_{2n} F_3$$

are compatible as n varies. We overcome the difficulty in the following manner. Firstly we observe that it suffices to deal with the case $n = \infty$. Hence we need an S-map

$$\tau : BO\, F_3 \longrightarrow B\Sigma_\infty \int O_2 F_3$$

such that τ_* composed with the canonical

$$(B\pi_\infty)_* : H_*(B\Sigma_\infty \int O_2 F_3) \longrightarrow H_*(BO\, F_3)$$

is an isomorphism (mod 2 coefficients here). This τ will then be inserted into an S-map composition like that of (2.10).

Let $\{X_\gamma(n) \subset BO_{2n} F_3\}$ be a cofinal family of finite subcomplexes. Set $P_\gamma(n)$ equal to the set of S-homotopy classes of S-map $X_\gamma(n) \to B\Sigma_\infty \int O_2 F_3$ such that when we form, as in (2.10, the S-map

$$X_\gamma(n) \longrightarrow \bigvee_{1 \le k} \frac{BO_{2k} F_3}{BO_{2k-2} F_3}$$

it induces on $H_*(\_;Z/2)$ the "canonical" map. Here we have identified

$$H_*\left(\frac{BO_{2k}F_3}{BO_{2k-2}F_3};Z/2\right)$$

with a subgroup of $H_*(BO\,F_3;Z/2)$ (see [11;21]) after which "canonical" map is interpreted as the map induced by the inclusion $X_\gamma(n) \subset BO_{2n}F_3 \to BO\,F_3$. Then $P_\gamma(n)$ is finite. It is also non-empty because, by an application of (2.7) in homology,

$$X_\gamma(n) \subset BO_{2n}F_3 \xrightarrow{\text{transfer}} B\Sigma_n\int O_2F_3 \to B\Sigma_\infty\int O_2F_3$$

is in $P_\gamma(n)$. Hence we may choose

$$\tau \in \varprojlim_{\gamma,n} P_\gamma(n) \subset \varprojlim_{\gamma,n} \{X_\gamma(n);B\Sigma_\infty\int O_2F_3\}$$

$$\cong \{BOF_3,B\Sigma_\infty\int O_2F_3\}$$

and form ϕ_∞ as in (2.10).

§3 First connections between classifying spaces and cobordism.

3.1. Recall that there exist Conner-Floyd classes $c_k \in MU^{2k}(BU)$ and $p_k \in MSp^{4k}(BSp)$. These classes lift uniquely to give $c_k' \in MU^{2k}(\frac{BU}{BU(k-1)})$ and $p_k' \in MSp^{4k}(\frac{BSp}{BSp(k-1)})$. The decompositions of §2.12 have the following immediate corollary.

3.2 Theorem. (a) Define

$$\Phi_U(n): \{X,\frac{BU}{BU(n-1)}\} \to \prod_{n\le k} MU^{2k}(X)$$

by

$$\Phi_U(n)(f) = \prod_{n\le k} f^*(c_k').$$

Then $\Phi_U(n)$ is an isomorphism if $\dim X \le 4n$ and is onto if $\dim X = 4n+1$.

(b) Define

$$\Phi_{Sp}(n): \{X,\frac{BSp}{BSp(n-1)}\} \to \prod_{n\le k} MSp^{4k}(X)$$

by

$$\Phi_{Sp}(n)(f) = \prod_{n\le k} f^*(p_k').$$

Then $\Phi_{Sp}(n)$ is an isomorphism if $\dim X \le 8n+2$ and is onto if $\dim X = 8n+3$.

Proof: For the unitary case §2.12 implies that $\{X,\frac{BU}{BU(n-1)}\}$ is isomorphic to $\prod_{n\le k}\{X,MU(k)\}$ if $\dim X < \infty$. It is well-known that the map $\{X,MU(k)\} \to MU^{2k}(X)$ given by pulling back a class $c_k'' \in MU^{2k}(MU(k))$ is an isomorphism when $\dim X \le 4k$ and is onto if $\dim X = 4k+1$. However c_k' restricts to c_k'' under the natural map. The result now follows by induction on $\dim X$ to show that, modulo lower skeletal filtration, $\Phi_U(n)$ is equal to the map induced by the c_k'''s. This completes part (a). Part (b) is entirely similar.

3.3. There is, of course, a real analogue to 3.2. However we will need a little more preparation in order to state the result. This is because we must recall the relationship between MO^* and $H^*(\_;Z/2)$ [25] in order to exhibit a $(4k-2)$-equivalence between $\frac{BO(2k)}{BO(2k-2)}$ and $MO(2k-1) \times MO(2k)$, a product of Thom spaces.

Let $d(\ell)$ equal the number of non-dyadic partitions of ℓ. That is, partitions $\ell = \Sigma a_i$ where $0 < a_i \ne 2^t-1$ for any t.

3.4 Theorem. (a) There is a $(4k-2)$-equivalence between $\frac{BO(2k)}{BO(2k-2)}$ and

$$\left(\prod_{h=0}^{2k-1} K(Z/2,2k-1+h)^{d(h)}\right) \times \left(\prod_{\ell=0}^{2k} K(Z/2,2k+\ell)^{d(\ell)}\right).$$

(b) There is a 2-local $(4n-2)$-equivalence between $\frac{BO_{2n}F_3}{BO_{2n-2}F_3}$ and a product of Eilenberg-Maclane spaces

$$\prod_{\varepsilon,\ell,q,J} K(Z/2,2n+\ell-\varepsilon+J)^{d(\ell)e(2n-q-\varepsilon,J)}.$$

The product is taken over $\varepsilon = 0$ or 1, $q \ge 0$ and $\ell-\varepsilon+J \le 2n-3$. Also $e(t,J)$ equals the number of partitions of the form

$$J+t = \sum_{i=1}^{k} (k_i+1) \qquad \text{with} \qquad 0 \le k_n \ne k_v$$

if $u \neq v$. Furthermore $e(t,J)$ and $d(\ell)$ are defined to be zero if $t < 0$ or $\ell < 0$ respectively and $d(0) = 1 = e(0,0)$.

3.5. The results of [25] show that the $(4k-2)$-skeleton of $\frac{BO(2k)}{BO(2k-2)}$ is homotopy equivalent to that of $MO(2k-1) \times MO(2k)$. The proof of Theorem 3.4 is accomplished by the method of [25]. For (b) one needs the mod 2 homology of BOF_3 which is described in [11].

In $MO^j(MO(j))$ there exists a canonical class, v_j. However, by §3.4 and the above remarks, we may lift (v_{2k-1}, v_{2k}) to give a class

$$u_k \in (MO^{2k-1} \times MO^{2k})\left(\frac{BO(2k)}{BO(2k-2)}\right)$$

which may be lifted back to give

$$u_k' \in (MO^{2k-1} \times MO^{2k})\left(\frac{BO}{BO(2k-2)}\right) .$$

The real analogue of §3.2, which admits a similar proof, is as follows.

3.6 Theorem. Define

$$\Phi_0(n)\colon \{X, \frac{BO}{BO(2n-2)}\} \to \prod_{2n-1 \leq t} MO^t(X)$$

by

$$\Phi_0(n)(f) = \prod_{n \leq k} f^*(u_k') .$$

Then $\Phi_0(n)$ is an isomorphism if $\dim X \leq 4n-3$ and is onto if $\dim X = 4n-2$.

§4 Computations in $\pi_j^s(BU)$ and some problems.

4.1. We have a spectral sequence

$$E^2_{p,q} = \tilde{H}_p(BU; \pi_q^s(S^0)) \Longrightarrow \pi^s_{p+q}(BU)$$

$(d_r\colon E^r_{p,q} \to E^r_{p-r,q+r-1})$. The associated filtration is

$$\underline{(0)} = F_{-1,k} \subset F_{0,k} \subset F_{1,k} \subset \cdots \subset F_{k,k} = \pi_k^s(BU)$$

$$E^\infty_{m,k-m} = \frac{F_{m,k}}{F_{m-1,k}} .$$

If $a_1, a_2 \in H_*(BU)$ we have two products, $a_1 * a_2$ and $a_1 a_2$, which are induced by the tensor product and Whitney sum respectively. The differentials behave as derivations with respect to both products. However §2.12, the differentials must respect the filtration of $H_*(BU)$ by $H_*(BU(n))$'s. The spectral sequence is a module over $\pi_*^S(S^0)$ and finally from [18] we know the behaviour of the differentials on $\tilde{H}_p(BU(1);\pi_q^S(S^0))$ (modulo odd torsion) through the range $p+q \leq 19$. This data forces the behaviour of the spectral sequence in low total degree. A routine calculation yields the following results. Further details of the generators are to be found in [20].

4.2 Theorem. The following table gives the value of $\pi_j^S(BU)/(\text{odd torsion})$ for $j \leq 10$ (zero groups are omitted).

j	$\pi_j^S(BU)/(\text{odd torsion})$
2	Z
4	$2Z$
5	$Z/2$
6	$3Z$
7	$Z/2$
8	$5Z \oplus Z/2$
9	$Z/8 \oplus Z/2$
10	$7Z \oplus Z/2$

4.3. The splittings of §§2.14-15 together with §3.4 combine to yield new elements in $\pi_*^S(BO)$ and $\pi_*^S(BOF_3)$. The precise results are as follows.

4.4 Theorem. There is a decomposition of stable homotopy groups

$$\pi_j^S(BO(2k)) \cong \pi_j^S(BO(2k-2)) \oplus B_j(k)$$

for all j,k. If $j < 4k-2$ then $B_j(k)$ contains the direct sum of $\beta_j(k)$ copies of $Z/2$ where

$$\beta_j(k) = d(j-2k+1)+d(j-2k)$$

and $d(\_)$ is as in §§3.3-3.4. These $Z/2$'s are detected by the mod 2 Hurewicz map.

4.5 Theorem. There is a decomposition of 2-localised stable homotopy groups

$$\pi_j^S(BO_{2n}F_3)_{(2)} \cong \pi_j^S(BO_{2n-2}F_3)_{(2)} \oplus C_j(n)$$

for all j,n. If $j < 4n-2$ then $C_j(n)$ contains the direct sum of $\gamma_j(n)$ copies of $Z/2$ where

$$\gamma_j(n) = \begin{cases} 0 & \text{if } j < 2n-1 \\ \sum\limits_{\varepsilon,\ell,q,J} d(\ell)e(2n-q-\varepsilon,J) & \text{if } 2n-1 \le j = 2n+\ell-\varepsilon+J \le 4n-3 \end{cases} .$$

Here $d(\_)$, $e(\_,\_)$ and the sum over ε,ℓ,q,J, are all as in 3.5(b). Each of these $Z/2$'s is detected by the mod 2 Hurewicz homomorphism.

4.6 Problems

4.6.1. BOF_3 has the same stable homotopy type as JO (see [21]) at the prime two. JO is a factor in SG, the one-component of $QS^0 = \varinjlim_n \Omega^n S^n$. Since SG is an infinite loopspace there is a homomorphism from its stable homotopy to its homotopy (induced by the structure map $d: QSG \to SG$). Thus we may form a homomorphism

$$\pi_j^s(BOF_3)_{(2)} \to \pi_j^s(SG)_{(2)} \xrightarrow{d_\#} \pi_j(SG)_{(2)} = \pi_j^s(S^0).$$

Thus Theorem 4.4 supplies an infinite number of $Z/2$'s mapping into the stable stems. Are any of the images non-zero?

4.6.2. The multiplication in KU- and KO-theory gives rise to maps as follows.

$$BO \wedge 0 \to 0$$

$$BO \wedge Sp \to Sp$$

$$BSp \wedge 0 \to Sp$$

$$BSp \wedge Sp \to 0$$

$$BU \wedge U \to U$$

In stable homotopy we obtain pairings like

$$\pi_j^s(BO) \otimes \pi_k^s(0) \to \pi_{j+k}^s(0).$$

These pairings are highly non-trivial on ordinary homotopy groups. Is it possible to construct infinite families of $Z/2$'s in $\pi_*^s(0)$ by successive pairings with the elements constructed in §4.4? What about the other four pairings?

4.6.3. Compute a range of the groups $\pi_j^s(BO)$ and $\pi_j^s(BSp)$.

4.6.4 Conjecture. Under the operations of Whitney sum and tensor product $\pi_2(BU(1))$ and $\pi_*(MU)$ generate $\pi_*^S(BU)$/torsion). This is true in dimensions ≤ 10. Here $\pi_*(MU)$ is embedded in $\pi_*^S(BU)$ by means of §§2.12, 3.2(a).

§5 A description of the classical cobordism theories

5.1. Let $\beta \in \pi_2(BU)$ be a generator. The Whitney sum, $\beta \oplus 1_{BU}: S^2 \times BU \to BU$, yields a map

(5.2) $$\varepsilon: \Sigma^4 BU \to \Sigma^2 BU$$

by means of the Hopf construction (suspended once). Also the Hopf construction applied to $1_{BU} \oplus \beta \oplus 1_{BU}$ yields a map

(5.3) $$m: \Sigma^4 BU \wedge BU \to \Sigma^2 BU.$$

Define a ring

(5.4) $$AU^0(X) = \varinjlim_N \{\Sigma^{2N} X, BU\}$$

where the limit is taken over composition with ε of (5.2) and multiplication is induced by m of (5.3) in the obvious manner. $AU^0(X)$ is a commutative ring because Whitney sum is homotopy commutative and homotopy associative.

A natural homomorphism of rings

(5.5) $$\Phi_U: AU^0(X) \to MU^{2*}(X) = \varinjlim_N \prod_{-N}^{\infty} MU^{2k}(X)$$

is defined as follows. If $x \in AU^0(X)$ is represented by $f \in \{\Sigma^{2N}X, BU\}$ set $\Phi_U(x)$ equal to the $2N$-th desuspension (in MU-theory) of $\prod_{1\leq k} f^*(c_k)$ where c_k is the k-th Conner-Floyd class of §3.1. Φ_U is independent of the choice of f. This is a simple calculation based on the fact that the total Conner-Floyd class, $C = c_0+c_1+\ldots \in MU^{2*}(BU)$, is a genus. That is, the pull-back of C under Whitney sum, $BU \times BU \to BU$, is $C \otimes C$.

Replacing BU by BSp, β by the generator of $\pi_4(BSp)$ and c_k by p_k of §3.1 we obtain

$$ASp^0(X) = \varinjlim_N \{\Sigma^{4N}X, BSp\}$$

(5.6) and

$$\Phi_{Sp}: ASp^0(X) \to MSp^{4*}(X).$$

Replacing BU by BO, β by the generator of $\pi_1(B)$ and c_k by the images of the classes u_k' of §3.5 we obtain

$$AO^0(X) = \varinjlim_N \{\Sigma^N X, BO\}$$

(5.7) and

$$\Phi_0: AO^0(X) \to MO^*(X).$$

<u>5.8 Theorem</u>. When $\dim X < \infty$ the ring homomorphisms Φ_U, Φ_{Sp} and Φ_0 of (5.5)-(5.7) are isomorphisms.

<u>Proof</u>: I will give a sketch of the symplectic case. This is the simplest, but it illustrates the method of proof which is common to all three cases. Firstly we observe that Φ_{Sp} is onto. This is because any $x \in MSp^{4*}(X)$ lies in some group

$$\prod_{n \leq k} MSp^{4k}(\Sigma^{4N+4n}X)$$

with $\dim X + 4N \leq 4n+2$ for n large enough. Hence by §3.2(b) $x \in \operatorname{im}(\Phi_{Sp})$.

Now suppose that $y \in \{\Sigma^{4N}X, BSp\}$ represents a class in $\ker \Phi_{Sp}$. We may suppose $N = 0$ and $y \in \{X, BSp(n)\} \cong \{X, MSp(n)\} \oplus \{X, BSp(n-1)\}$. This last splitting results from §2.12. Similarly

(5.9) $$\varepsilon_{\#}(y) \in \{\Sigma^4 X, MSp(n+1)\} \oplus \{\Sigma^4 X, BSp(n)\}$$

because the decompositions of §2.12 respect filtration by $BSp(n)$'s and $S^4 \times BSp(n)$ maps by $\beta \oplus 1_{BSp}$ to $BSp(n+1)$. Consider the two components of $\varepsilon_{\#}(y)$ in (5.9). For the first we have

$$\dim X - 8n - 2 > \dim(\Sigma^4 X) - (8n+8) - 2.$$

This component has image in $MSp^{4n}(X)$ and is mapped injectively if $\dim(\Sigma^4 X) - 8(n+1) - 2$ is negative. Hence, by induction on the quantity (associated with $\{Y, BSp(s)\}$)

dim Y-8s-2, we may assume that the first component of $\varepsilon_\#(y)$ in (5.9) becomes zero in the direct limit. For the second component we will find another numerical quantity on which to base an induction. Let W_t denote the t-skeleton of W. The S-map, obtained from ε,

$$\Sigma^4 BSp(n) \to MSp(n+1) \vee BSp(n) \to SBp(n)$$

induces zero on $H_*(\_;Z)$. Hence it may be deformed to send $\Sigma^4(BSp(n)_{4u})$ into $BSp(n)_{4u}$. However the obstruction to a further deformation which sends $\Sigma^4(BSp(n)_{4u})$ into $BSp(n)_{4u-4}$ vanishes because $\pi_4^S(S^0)$ is zero. Thus if y is represented by an S-map into $BSp(n)_{4u}$ the second component in (5.9) is represented by an S-map into $BSp(n)_{4u-4}$. Therefore induction on the skeletal filtration of representatives shows that the second component of $\varepsilon_\#(y)$ becomes zero in the direct limit. Hence y represents zero and Φ_{Sp} is injective.

More details and the proofs of the other cases are to be found in [20, Parts II §2, III §3].

§6 The X(T)-spectrum

6.1: In this section X will be a homotopy commutative and homotopy associative H-space. Also $T = \{B_i \in \pi_{n_i}^S(X);\ 1 \le i \le t\}$ will be a set of stable homotopy classes. Set $b = B_1 B_2 \ldots B_t \in \pi_N^S(X)$, the "product" of the B_i's. From this data may be constructed a periodic ring spectrum, X(T), of period N. Find the smallest interger $u \ge 0$ such that b may be represented by a map

$$b: S^{(u+1)N} \to \Sigma^{uN} X. \tag{6.2}$$

Then $\Sigma^N b$ will become the unit of X(T). Set $M = (1+u)N$. Now for the spectrum. Put

$$X(T)_{kN} = \Sigma^M X \quad \text{for} \quad k \ge 1.$$

Define structure maps

$$\varepsilon: \Sigma^N X(T)_{kN} \to X(T)_{(k+1)N}$$

by means of the composition

(6.3)

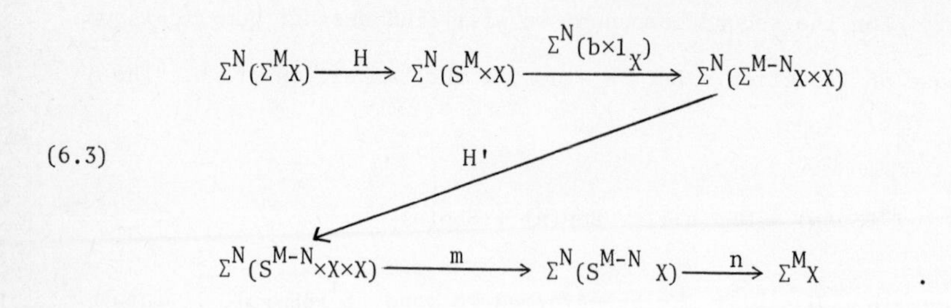

$$\Sigma^N(\Sigma^M X) \xrightarrow{H} \Sigma^N(S^M \times X) \xrightarrow{\Sigma^N(b\times 1_X)} \Sigma^N(\Sigma^{M-N} X\times X) \xrightarrow{H'} \Sigma^N(S^{M-N}\times X\times X) \xrightarrow{m} \Sigma^N(S^{M-N}\ X) \xrightarrow{n} \Sigma^M X \quad .$$

In (6.3) H and H' are given by Hopf constructions, m is induced by the H-space structure in X and n is the canonical collapse. A constructon similar to (6.3) in which $1_X \times b \times 1_X$ replaces $b \times 1_X$ defines a pairing

$$X(T)_{kN} \wedge X(T)_{\ell N} \to X(T)_{(k+\ell)N} \quad .$$

The following is straightforward [20, Part III].

<u>6.4 Theorem</u>. The spectrum, $X(T)$, described above is a commutative ring spectrum. If $\beta \in \pi_2(BU)$ is a generator then $BU(\beta)^0(X) \cong AU^0(X)$ of (5.4). If $b \in \pi_4(BSp)$ is a generator then $BSp(b)^0(X) \cong ASp^0(X)$ of (5.6). If $\eta \in \pi_1(BO)$ is a generator then $BO(\eta)^0(X) \cong AO^0(X)$ of (5.7). (These examples will be called the AU-, ASp- and AO-spectrum respectively.)

Here are some further examples of $X(T)$-spectra which seem to be of interest.

<u>6.5 Algebraic cobordism of A</u>. Let A be any ring with unit then set $X = BGLA^+$ [26]. For $T \subset \pi_*^S(BGLA^+)$ the resulting cohomology theory is called the *algebraic cobordism of* A *associated with* T.

Of particular interest is the case $A = Z$, the integers, and T consists of a generator of $\pi_1(BGLZ^+) \cong Z/2$. Denote this special example by AZ-theory. The inclusion $Z \subset R$ induces a map of ring spectra

$$r\colon AZ^* \to AO^* \quad .$$

Both AZ- and AO-theory are Z/2-vector spaces because the class, b, of (6.2) and (6.3) is of order 2. In §7 it will be shown that

$$r: AZ^0(X) \to AO^0(X)$$

is onto if $\dim X \leq \infty$ and that when X is a sphere r is not an isomorphism.

6.6. We may replace BU in (6.4) by $BU\Lambda$, the classifying space for $KU(\_;\Lambda)$-theory. For example if $\Lambda = Z_{(p)}$ (the integers localised at p) or $\Lambda = \hat{Z}_p$ (the p-adics) then the resulting cohomology theories will be isomorphic to $MUZ^{2*}_{(p)}$ and $M\hat{UZ}^{2*}_p$ respectively. The proof for these cases is the same as that of (5.8).

6.7. If A is a commutative ring with unit we may set $X = BGLA^{\wedge}_{et}$, the profinite completion away from the characteristics of A of the étale homotopy type of the classifying space for GLA. Étale homotopy types are in general only pro-spaces whose stable homotopy theory is not, to my knowledge, established. However, in favourable circumstances, techniques of Sullivan and others [12;13] may be used to construct from an étale homotopy type a single space. It is this *single space* [20, Part IV] that I have in mind for $BGLA^{\wedge}_{et}$. For example, when $A = C$, the complex numbers, or A is an algebraically closed field of finite characteristic this construction may be performed. More of this in §9.

6.8. Set $X = BU^{\otimes}$, the classifying space for the group of special units $1+\tilde{K}U^0(\_)$ in unitary K-theory. If T consists of the generator of $\pi_2(BU^{\otimes})$ then $BU^{\otimes}(T)$-theory turns out to contain a factor which is (c.f. §6.9) isomorphic to periodic unitary K-theory.

6.9. Set $X = CP^{\infty}$ and let T consist of the generator of $\pi_2(CP^{\infty})$. The H-space CP^{∞} classifies the group of line bundles under tensor product. Surprisingly this cohomology theory equals KU-theory on finite dimensional spaces. This will be proved below in §6.10-6.13.

This answers a question of D.S. Kahn [28].

6.10. Let $T \in \pi_2(CP^{\infty})$ be as in §6.9. Form the homomorphism

$$(T.\_): \pi^S_j(CP^{\infty}) \to \pi^S_{j+2}(CP^{\infty})$$

given by "adding T" by means of the H-space structure on $CP^{\infty} = K(Z,2)$.

The homotopy groups of the spectrum of §6.9 are given by

$$\pi_j(CP^{\infty}(T)) = \varinjlim_k \pi^S_{j+2k}(CP^{\infty}) \tag{6.11}$$

where the limit in (6.12) is over iterated compositions of $(T.\_)$. If this is to be $\pi_j(BU)$ then (6.12) must be torsion free. Remarkably it is.

<u>6.12.Theorem</u>. Let $y \in \pi_j^S(CP^\infty)$ be a torsion element. Then there exists $0 \leq k \in Z$ such that

$$0 = T^k y \in \pi_{j+2k}^S(CP^\infty).$$

<u>Proof</u>: Factor the composition $(T.\_)^k$ as follows.

$$\pi_j^S(CP^\infty) \xrightarrow{i_\#} \pi_j^S(BU) \xrightarrow{\beta^k.\_} \pi_{j+2k}^S(BU) \xrightarrow{(Bdet)_\#} \pi_{j+2k}^S(CP^\infty).$$

Here $i\colon CP^\infty = BU(1) \to BU$ is the natural map and $\beta \in \pi_2(BU)$ is a generator, $\beta = i_\#(T)$. Also $B\,det\colon BU \to CP^\infty$ is the H-map induced by the determinant. It is well-known that $(Bdet)\circ i = 1$ so the above composition is indeed $(T.\_)^k$. However, we know from Theorem 5.8 that

$$\varinjlim_k \pi_{2k+j}^S(BU) = \pi_j(\underline{MU}^{2*})$$

which is torsion free. Here the limit is taken over $(\beta.\_)$. Hence there exists a k such that

$$\beta^k i_\#(y) = 0$$

and hence

$$T^k\, y = (Bdet)_\# \beta^k i_\#(y) = 0$$

as required.

<u>6.13 Theorem</u>. $CP^\infty(T)$ is equivalent to the periodic $\underline{BU}$-spectrum. Equivalently there is an equivalence of infinite loopspaces

$$BU \simeq \varinjlim_n \varinjlim_k \Omega^{2n+2k}\Sigma^{2k}CP^\infty$$

in which the limit is taken over composition with $\varepsilon\colon \Sigma^4 CP^\infty \to \Sigma^2 CP^\infty$, the structure map of $CP^\infty(T)$.

<u>Proof</u>: Firstly one checks that pulling back the reduced Hopf bundle

$$x \in KU^0(CP^\infty) \cong Z[[x]]$$

gives a natural transformation

$$\phi\colon \varinjlim_k \{\Sigma^{2k}\_, CP^\infty\} \to KU^0(\_).$$

Next one notes that the diagram

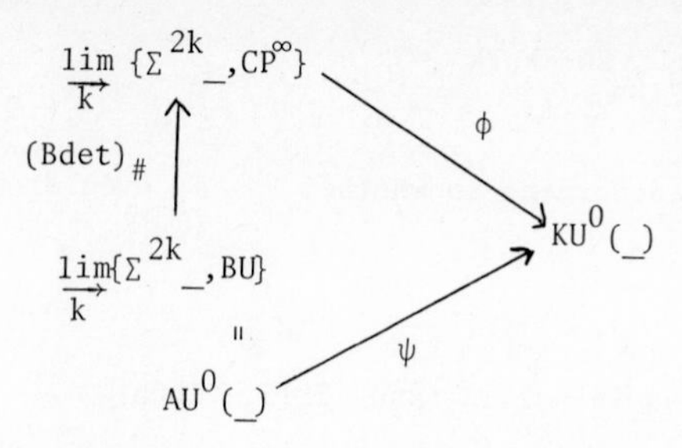

commutes where ψ is obtained by pulling back the determinant (or total γ-class) $\sum_{0\leq i} \gamma^i \in KU^0(BU)$.

Now define the Conner-Floyd homomorphism $\lambda: MU^{2*}(\_) \to KU^0(\_)$ by pulling back the universal Thom class

$$\sum_{0\leq n} \Lambda_n \in \prod_n \widetilde{KU}^0(MU(n)).$$

Since the Thom class, $\Lambda_n \in \widetilde{KU}^0(MU(n))$, restricts to the n-th γ-class

$$\gamma^n \in KU^0(BU(n))$$

a computation (which requires knowledge of the S-maps of Theorem 2.12 in KU_*_theory) shows that

$$\lambda \circ \Phi_U = \phi$$

where Φ_U is the isomorphism $AU^0(\_) \cong MU^{2*}(\_)$. The first MU-theory Conner-Floyd class gives a homomorphism $c_1: KU^0(\_) \to MU^{2*}(\_)$ and an easy calculation (in the universal case) shows that $\lambda \circ c_1 = -1$.

Hence λ, ψ and ϕ are onto maps on homotopy groups. By §6.12 the homotopy of $CP^{\infty}(T)$ is torsion free. Now a rational calculation shows that

$$\text{rank } \pi_j(\varinjlim_n \varinjlim_k \Omega^{2n+2k}\Sigma^{2k}CP^{\infty}) = \begin{cases} 1 & j \text{ even} \\ 0 & j \text{ odd} \end{cases}.$$

This rational calculation can be accomplished by observing that

$$\varinjlim_k \Omega^{2k}\Sigma^{2k}CP^{\infty} \qquad \text{and} \qquad SP^{\infty}CP^{\infty}$$

are equal rationally. Here $SP^{\infty}CP^{\infty}$ is the infinite symmetric product. By [9] rationally $SP^{\infty}CP^{\infty}$ equals $\prod_{1\leq i} K(Q,2i)$. It is now easy to evaluate

$$\varinjlim_n \varinjlim_k H_j(\Omega^{2n+2k}\Sigma^{2k}CP^\infty;Q).$$

Since $\pi_j(BU = Z$ or 0 according to whether j is even or odd the result follows because ϕ is onto.

<u>6.14 Problem</u>. Let BOF_3 be as in 2.13 and 2.15. Then

$$H^*(BOF_3;Z/2) = H^*(BO;Z/2) \otimes H^*(SO;Z/2).$$

Also $\pi_1^s(BOF_3) \cong Z/2 \oplus Z/2$. Let T consist of any element in $\pi_1^s(BOF_3)_{(2)}$ which maps non-trivially under the Brauer lifting map $BOF_3^+ \to BO$ (see D. Quillen: The Adams Conjecture, Topology 10 (1971) 67-80). The cohomology of the spectrum $BOF_3^+(T)$ looks like that of the smash product of the suspension spectrum of SO with the AO-spectrum.

Are these theories equal?

§7 AZ-theory

<u>7.0</u>. The homomorphism $Z \subset R$ induces on H-map $r\colon BGLZ^+ \to BO \simeq BGLR$ which induce and isomorphism on π_1. Hence we obtain a natural ring homomorphism (c.f. §6.5)

$$r\colon AZ^* \to AO^*. \tag{7.1}$$

By §5.8 we have a ring homomorphism

$$\Phi_0\colon AO^0(X) \to MO^*(X). \tag{7.2}$$

Composing (7.1) and (7.2) we obtain a homomorphism from $AZ^0(X)$, a large $Z/2$-vector space, to $MO^*(X)$ whose identity is known [25].

<u>7.3 Theorem</u>. The composition $\Phi_0 \circ r$ is onto if $\dim X < \infty$. It has non-trivial kernel when X is any sphere.

Proof: A non-zero element in $\ker(\Phi_0 \circ r)$ is constructed as follows. From [16] we know $\pi_3(BGLZ^+) \cong Z/48$. From any X there is an exponential map

$$(7.4) \qquad \nu : [X, BGLZ^+] \to AZ^0(X)$$

obtained by sending f to $1+\bar{\nu}(f)$ where $\bar{\nu}(f) \in AZ^0(X)$ is represented by the suspension of f, $\Sigma f \in [\Sigma X, AZ_1]$. When $X = S^3$ the generator has an image under ν which may be detected by means of the Hurewicz homomorphism $\pi_3(\underline{AZ}) \to H_3(\underline{AZ}; Z/2)$. The diagram

$$\begin{array}{ccc} \pi_3(BGLZ^+) & \xrightarrow{\nu} & AZ^0(S^3) \\ \downarrow r_\# & & \downarrow r \\ \pi_3(BO) & \xrightarrow{\nu} & AO^0(S^3) \end{array}$$

commutes thus showing that $r \circ \nu$ is trivial on $\pi_3(BGLZ^+)$ since $\pi_3(BO)$ is zero. Since AZ- and AO-theory are both periodic (of period one) we see that $\Phi_0 \circ r$ has non-trivial kernel for $X = S^n$ $(n \geq 0)$.

To show that r is onto for finite dimensional X we recall that the proof of §5.8 (in the real case) shows that generating classes in $MO^*(X)$ may be represented by S-maps of the form

$$f: \Sigma^N X \to \frac{BO(2k)}{BO(2k-2)}$$

with $\dim X + N < 4k-3$. From §3.4(b) we note that the $(4k-3)$-skeleton of $\frac{BO(2k)}{BO(2k-2)}$ may be considered as a split factor in the $(4k-3)$-skeleton of $\frac{BO_{2k}F_3}{BO_{2k-2}F_3}$. Here and for the rest of the proof we are working 2-locally.

Now O_2F_3 is generated by unimodular matrices

$$\begin{pmatrix} 0 & 1 \\ -1 & 0 \end{pmatrix} \quad \text{and} \quad \begin{pmatrix} 1 & 0 \\ 0 & -1 \end{pmatrix}.$$

Hence we have $O_2F_3 \subset GL_2Z \subset GL_2R$. We obtain a diagram of maps

(7.5)

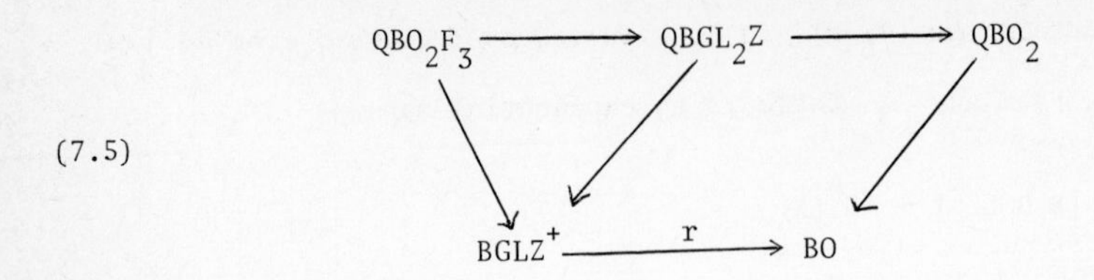

In (7.5) $QX = \varinjlim_n \Omega^n\Sigma^n X$ and the λ_i are the infinite loop maps induced by the canonical maps such as $BO_2F_3 \to BGL_2Z \to BGLZ^+$. Also there is an equivalence, by a result of Barratt-Priddy-Quillen et al.,

$$B\Sigma_\infty \int O_2F_3{}^+ \simeq QBO_2F_3 \ .$$

Now the S-map of §2.15 (proof) has an adjoint

$$\tau' \colon BOF_3 \to QB\Sigma_\infty \int O_2F_3{}^+ \ .$$

Hence we may form a composite

$$\text{(7.6)} \qquad \tau'' \colon BOF_3 \xrightarrow{\tau'} QQBO_2F_3 \to QBO_2F_3$$

in which the second map is the structure map of the free infinite loop space functor, Q. That is, the map induced by evaluation

$$\Omega^n\Sigma^n\Omega^n\Sigma^n \to \Omega^n\Sigma^n.$$

Choose an S-map

$$f' \colon \Sigma^N X \longrightarrow \frac{BO_{2k}F_3}{BO_{2k-2}F_3}$$

which, when considered as landing in the (4k-3)-skeleton, "projects" to the given map f. Here the "projection" is in the sense mentioned above whereby the (4k-3)-skeleton of $\frac{BO(2k)}{BO(2k-2)}$ is considered as a factor in that of $\frac{BO_{2k}F_3}{BO_{2k-2}F_3}$. Now form the composition of S-maps

$$\text{(7.7)} \qquad \Sigma^N X \xrightarrow{f'} \frac{BO_{2k}F_3}{BO_{2k-2}F_3} \to BOF_3 \xrightarrow{\tau''} QBO_2F_3 \xrightarrow{\lambda_3} BGLZ^+ \ .$$

Here τ'' and λ_3 are as in (7.6) and (7.5) while the second S-map is the "inclusion" induced by the splitting, Φ_∞, of §2.15. A mod 2 homology computation of the composition of (7.7) with r shows that (7.7) represents a class in $AZ^0(X)$

whose image under r is the same class as that represented by f. Note that the induced map in $H_*(\_;Z/2)$ characterises an AO^0-theory class in view of the isomorphism Φ_0 of (5.7).

The proof sketched above is elaborated in [22].

§8 Classical cobordism constructions in AU-theory

8.1. In this section the familiar MU-theory phenomena:- Adams operations, idempotents, Landweber-Novikov operations, the complexification and Conner-Floyd homomorphisms, the Thom isomorphism and the Pontrjagin-Thom construction -- will be described in terms of the AU-spectrum.

Firstly let us record a useful lemma.

8.2. Let Λ be as in §6.6 and let h^* be a multiplicative cohomology theory. Suppose that

$$\alpha: KU^0(\_;\Lambda) \to h^*(\_)$$

is a natural exponential map such that $\alpha(\beta) \in h^*(S^2)$ projects to the suspension class $\sigma \in h^2(S^2)$. Here β is as in §5.1 a "projection" means project onto reduced h^*-theory.

Then α induces a natural ring homomorphism

$$\hat{\alpha}: AU\Lambda^0(\_) \to h^*(\_).$$

If x is represented by $f \in [\Sigma^{2N}X, AU\Lambda_{2N}]$ then $\hat{\alpha}(x) = f^*(\sigma\otimes\alpha) \in h^*(\Sigma^{2N}X) \cong h^{*-2N}(X)$.

8.3 Example. Define $\nu: KU^0(\_;\Lambda) \to AU\Lambda^0(\_)$ by setting $\nu(f)-1 \in AU\Lambda^0(X)$ equal to the class represented by the double suspension of f, $\Sigma^2 f: \Sigma^2 X \to AU\Lambda_2$. It is clear that $\nu(\beta)-1$ is the double suspension of the unit for AU -theory and it is straightforward to check the exponential property. Of course $\hat{\nu}$ is just the identity.

8.4 The Conner-Floyd homomorphism. Let $\alpha = \gamma^0+\gamma^1+\gamma^2+\ldots$, the total γ-class in KU-theory. That is, α is induced by the determinant homomorphism $\det: U \to S'$. We obtain

$$\hat{\alpha}: AU^0(X) \to KU^0(X).$$

To see that this is essentially (i.e. modulo a suitable filtration) equal to the Conner-Floyd homomorphism [8] we need only observe that the latter is induced by the Thom class in $KU^0(MU(n))$ which restricts to the n-th γ-class in

$$KU^0(\frac{BU}{BU(n-1)}) \subset KU^0(BU).$$

Here we have identified $AU^0(X)$ with $MU^{2*}(X)$ by means of §5.8.

In terms of the model for KU-theory given in §6.9 $\hat{\alpha}$ is just the induced map of ring spectra

(8.5) $$\det: BU(\beta) \to CP^\infty(\beta)$$

given by the determinant.

<u>8.6 Theorem [8]</u>. When $\dim X < \infty$ the map (8.5) induces an isomorphism

$$MU^{2*}(X) \underset{\pi_* MU}{\otimes} Z \to KU^0(X).$$

<u>Proof:</u> By §5.8 and the discussion of §8.4 we are lead to consider the homomorphism

$$\det_{\#}: \varinjlim_N [\Sigma^{2N}X, \Sigma^2 BU] \to \varinjlim_N [\Sigma^{2N}X, \Sigma^2 CP^\infty].$$

Since $\det: BU \to CP^\infty$ is a split surjection $\det_{\#}$ is onto, being a direct limit of split surjections. Since $\pi_* MU$ is mapped to $KU^0(pt) = Z$ the ring epimorphism $\det_{\#}$ induces a surjection

(8.7) $$MU^{2*}(X) \underset{\pi_* MU}{\otimes} Z \to KU^0(X).$$

If X has only even dimensional cells then (8.7) is a surjection between free abelian groups of the same rank and hence is an isomorphism. Now pass to the general case by means of a "geometric resolution" of X by spaces having only even dimensional cells (c.f. [8]).

<u>8.8 Corollary</u> [14;24]. The Boardman-Hurewicz homomorphism

$$\pi_*(MU) \to KU_*(MU)$$

is a split injection.

Proof: In [27] it is shown how to deduce the Hattori-Stong result from the isomorphism of §8.6.

8.9 Adams operations. Let Λ be a ring, as in §6.6, which contains $1/k$ for some $k \in Z$. Then $1/k\ \psi^k\colon BU\Lambda \to BU\Lambda$ is an H-map which is the identity on π_2. Here $\psi^k\colon KU^0(\_;\Lambda) \to KU^0(\_;\Lambda)$ is the Adams operation. If ν is as in Example 8.3 the composite $\nu_0(1/k\ \psi^k)$ induces, by §8.2, a ring endomorphism

(8.10) $$\psi^k\colon AU\Lambda^* \to AU\Lambda^*.$$

8.11 Theorem. The operation of (8.10) satisfies the following properties.

(a) $\psi^k \circ \psi^\ell = \psi^{k\ell}$ when defined.

(b) The endomorphism of $MU\Lambda^{2*}(S^{2N})$ given by $\Phi_U \circ \psi^k \circ \Phi_U^{-1}$ is equal to multiplication by k^{N-t} on $MU\Lambda^{2t}(S^{2N})$. Here Φ_U is as in §5.8.

(c) Let $w \in AU\Lambda^0(CP^T)$ satisfy $\Phi_U(w) = c_1(y) \in MU\Lambda^2(CP^T)$ the first Conner-Floyd class of the Hopf bundle, y.)

Then

$$\Phi_U(\psi^k(w)) = \prod_{i\leq j} c_j(\frac{y^k}{k}) \in MU\Lambda^{2*}(CP^T).$$

Here c_j is the j-th Conner-Floyd class.

Proof: These properties follow easily from well-known properties of the K-theory Adams operations.

8.12. If $\dim X < \infty$ we may define

$$\psi^k\colon MU\Lambda^{2n}(X) \subset MU\Lambda^{2*}(X) \xrightarrow{\Phi_U \circ \psi^k \circ \Phi_U^{-1}} MU\Lambda^{2*}(X) \to MU\Lambda^{2n}(X)$$

in which the last map projects to dimension 2n. By periodicity of AU -theory we may extend this to a homomorphism of graded rings

(8.13) $$\psi^k\colon MU\Lambda^*(\_) \quad MU\Lambda^*(\_)\ .$$

8.13 Corollary. (Existence of cobordism Adams operations). The operation of (8.13) satisfies the following properties.

(a) $\psi^k \circ \psi^\ell = \psi^{k\ell}$ when defined.

(b) ψ^k is multiplication by k^{N-t} on $MU\Lambda^{2t}(S^{2N})$.

(c) $\psi^k(c_1(y)) = \frac{1}{k} c_1(y^k) \in MU\Lambda^2(CP^\infty)$.

<u>8.15 The idempotents of Adams</u>. For $d > 1$ let $R(d)$ be the ring of fractions a/b such that b contains no prime p with $p \equiv 1$ (d). Then the K-theory idempotent of Adams [2, p.93]

$$E_1 : BUR(d) \to BUR(d)$$

is the identity on π_2. By means of §8.2 applied to $\nu_0 E_1$, where ν is as in §8.3, we obtain a natural idempotent ring homomorphism

$$\varepsilon(d): AUR(d)^*(\_) \to AUR(d)^*(\_) . \tag{8.16}$$

By means of 5.8 this induces an idempotent in MUR(d)-theory which coincides with the idempotent of Adams [2,p.107].

It is important to note that this construction of Adams idempotent uses no knowledge of $\pi_*(MU)$, unlike the original proof (ibid) which uses the Hattori-Stong theorem.

<u>8.17 Theorem</u>. The idempotent of (8.16) satisfies the following properties.

(a) If $p \equiv 1$ (d) is a prime, $\varepsilon(d)$ induces an idempotent of $AU\hat{Z}_p^*(\_)$ satisfying

$$[\varepsilon(d)(f)]^d = \prod_{j=1}^{d} \psi^{\alpha_j}(f)$$

($f \in AU\hat{Z}_p^0(X)$). Here $\alpha_1, \alpha_2, \ldots, \alpha_d$ are distinct p-adic d-th roots of unity and ψ^{α_j} is the corresponding Adams operation of §8.10.

(b) If $\dim X < \infty$ $\varepsilon(d)$ induces (c.f. §8.12) an indepotent, ε, of $MUR(d)^*$ which is equal to that of Adams [2, p.107]. With p-adic coefficients and $p \equiv 1(d)$ we have

$$[\varepsilon(f)]^d = \prod_{j=1}^{d} \psi^{\alpha_j}(f)$$

($f \in MU\hat{Z}_p^{2n}(X)$). Here ψ^{α_j} is as in §8.14.

<u>Proof</u>: Part (b) follows from part (a) together with the fact that the Adams idempotent is characterised by its effect on $\pi_*(MUR(d))$. The formula in (a) follows

from the equation

$$E_1 = \frac{1}{d}\left(\sum_{j=1}^{d} \psi^{\alpha_j}\right) \qquad [2]$$

for the K-theory idempotent. The sum is given by Whitney sum in BUR(d) which induces the product in AUR(d)-theory.

8.18 The idempotent of Quillen. From the splitting principal an exponential map $\alpha: KU^0(\_;\Lambda) \to MU\Lambda^*(\_)$ is determined by

$$\alpha(y-1) \in MU\Lambda^*(CP^\infty) \cong \pi_*(MU\Lambda)[[x]].$$

Here y is the Hopf line bundle. Following [1,p.108] define

$$\text{mog } x = \log x - \frac{1}{d}\left(\sum_{j=1}^{d} \log(\alpha_j x)\right) \qquad (d > 1)$$

where α_j is as in §8.17. Then

$$\text{mog } x \in \pi_*(MUZ[\tfrac{1}{d}])[[x]]$$

and induces an endomorphism of $AUZ[\frac{1}{d}]^*$ which in turn induces Quillen's idempotent on $MUZ[\frac{1}{d}]^*$.

8.19 The complexification homomorphism. Consider the complexification map $c: BSp \to BU$. On π_4^s the generator $B \in \pi_4^s(BSp) \cong Z$ is sent to $x^2 - x*x \in \pi_4^s(BU)$ where x generates $\pi_2^s(BU) = Z$. From §4.2 it is easy to see that $\pi_4^s(BU) = Z \oplus Z$ generated by x^2 and $x*x$. Thus we get a commutative diagram of S-maps:

$$\begin{array}{ccc}
\Sigma^4(S^4 \times BSp) & \xrightarrow{\Sigma^4(B \oplus 1_{BSp})} & \Sigma^4 BSp \\
\downarrow{\scriptstyle \Sigma^4(1_{S^4} \times c)} & & \downarrow{\scriptstyle \Sigma^4 c} \\
\Sigma^4(S^4 \times BU) & \xrightarrow{\Sigma^4((x^2 - x*x) \oplus 1_{BU})} & \Sigma^4 BU
\end{array} \qquad (8.20)$$

Note that $\Sigma^4(x^2 \oplus 1_{BU})$ is the map which induces $\varepsilon^2: \Sigma^8 BU \to \Sigma^4 BU$ where ε is the structure map of the AU-spectrum. Also $x*x$ corresponds to

$$a_{11} \in \pi_2(MU) \subset MU^{2*}(S^0) \cong AU^0(S^0).$$

Thus we find that composition with c does not define a map $ASp^* \to AU^*$. But it does define a complexification homomorphism of the form

$$(8.21) \qquad c\colon ASp^0(X) \to AU^0(X)[1-a_{11}]^{-1}.$$

It should be remarked that we have not lost a great deal by inverting

$$1-a_{11} \in MU^{2*}(S^0)$$

since it is invertible in $\prod_{k=-\infty}^{\infty} MU^{2k}(\_)$-theory, which contains MU^{2*}.

By virtue of §5.8 we have the following result.

8.22 Proposition. Let $\pi\colon E \to X$ be a complex n-plane bundle. Let X be compact and let $Th(E)$ be the Thom space of E. Then there is a Thom class

$$\lambda_E \in \tilde{AU}^0(Th(E))$$

such that $\lambda_{E\oplus F} = \lambda_E\lambda_F$ and $\lambda(x) = \pi^*(x)\lambda_E$ defines an isomorphism

$$\lambda\colon AU^0(X) \xrightarrow{\cong} AU^0(Th(E)).$$

Furthermore if $n = 1$ and $\beta_E\colon Th(E) \to BU$ is the K-theory Thom class then λ_E is represented by $\Sigma^2\beta_E$.

8.23 Landweber-Novikov operations

For each finitely non-zero sequence of positive integers we have [1,p.9] a Conner-Floyd class

$$c_\alpha \in MU^{2|\alpha|}(BU) \qquad (\alpha = (\alpha_1,\alpha_2,\ldots);\ |\alpha| = \Sigma\alpha_i).$$

Then $C = \sum_\alpha c_\alpha$ is an exponential operation to which we may apply §8.2. We obtain the "super-total" Landweber-Novikov operation

$$S = \hat{C}\colon AU^0(X) \to AU^0(X) \cong MU^{2*}(X).$$

S behaves like $\sum_\alpha s_\alpha$, the "sum" of all the Landweber-Novikov operations. Since S is a ring homomorphism it may be characterised by its effect on the canonical class $w \in AU^0(CP^\infty)$ where $\Phi_U(w) \in MU^2(CP^\infty)$ is the canonical class. Now the value of C on the reduced Hopf bundle is $\sum_{i\geq 0} w^i$. Also $\Phi_U(S(w)) = f^*(r \otimes C)$ where $f\colon \Sigma^2CP^\infty \to \Sigma^2BU$ is the canonical map. Thus by [1,p.9] $\Phi_U(S(w)) = \sum_{i\geq 1} (\Phi_U(w))^i$.

We easily obtain the following result.

8.24 Theorem. The homomorphism, S, of §8.23 satisfies

(i) $S(w) = \sum_{i\geq 1} w^i$

(ii) Suppose that $E \to X$ is a complex n-plane bundle. Consider the diagram

$$\begin{array}{ccc} \widetilde{AU}^0(Th(E)) & \xrightarrow{S} & \widetilde{AU}^0(Th(E)) \\ \lambda \uparrow \cong & & \cong \uparrow \lambda \\ AU^0(X) & \longrightarrow & AU^0(X) \end{array}$$

then $\lambda^{-1}(S(\lambda(1))) = C(E)$ where λ and $Th(E)$ are as in §8.22.

8.25 Pontrjagin-Thom construction.

It is possible to distinguish unitary cobordism classes by means of the Pontrjagin-Thom construction based on Thom classes for AU-theory which are obtained entirely from the transfer-type constructions in homotopy theory.

We will look at this construction both from the geometric and from the cohomology-theoretic viewpoints.

Let $\Lambda_n \in AU^0(M(n))$ be the class represented by the S-map inclusion given by §§1.1, 2.12. A computation of

$$(\Lambda_n)_*: MU_{2*}(MU(n)) \to MU_{2*}(BU)$$

similar to the homology computations necessary for §2.12 show that Λ_n will serve as the universal Thom class for AU-theory and complex n-planes. It suffices to show that the MU^{2n}-component of Λ_n is the usual Thom class. Then we obtain the following result.

8.26 Theorem. Let M^{2n} be a closed stably almost complex manifold. Let $\nu: M^{2n} \to BU(k)$ classify the stable normal bundle and let $\sigma(\nu): Th(\nu) \to MU(k)$ be the "Thomification" of ν. Let $P(i): S^{2n+2k} \to Th(\nu)$ be the Pontrjagin-Thom map of the embedding $M^{2n} \subset R^{2n+2k}$.

Then the association

$$M^{2n} \to (\pi_{2n}(MU)\text{-component of } P(i)^*\sigma(\nu)^*\Lambda_k)$$

yields a bjection

$$\text{(unitary bordism classes in dimension 2n)} \Longleftrightarrow \pi_{2n}(MU).$$

8.27. Now let us continue with the Pontrjagin-Thom construction from the homotopy theory point of view. This is just a straightforward exercise in manipulating pairings in AU-theory.

Suppose that $f: M^{2n} \to N^{2n+2k}$ is a smooth embedding with complex normal bundle, ν. We have in mind the case $N = R^{2n+2k}$. Set

$$\Lambda(\nu) = \sigma(\nu)^*(\Lambda_k) \in \widetilde{AU}^0(Th(\nu)).$$

We have an exact sequence of bundles

$$0 \to \tau_M \to f^*\tau_N \to \nu \to 0$$

where τ_M and τ_N are the tangent bundles. Thus we have a slant pairing in AU-homology

$$(\Lambda(\nu)\backslash\_): \widetilde{AU}_0(Th(\nu)) \to AU_0(M) \tag{8.28}$$

which is an isomorphism. This is the dual of the Thom isomorphism. We also have the Kronecker pairing

$$\langle\Lambda(\nu),\_\rangle: \widetilde{AU}_0(Th(\ )) \to \pi_0(AU). \tag{8.29}$$

Thus we may compose the inverse of (8.28) with (8.29) to obtain

$$\lambda(f): AU_0(M) \to \pi_0(AU) \cong \pi_{2*}(MU). \tag{8.30}$$

When $N = R^{2n+2k}$ the image of the fundamental class $[M] \in MU_{2n}(M)$ has $\pi_2(MU)$-component equal to that of $P(i)^*\sigma(\nu)^*(\wedge_k)$ of §8.26. This is an exercise for the reader. Details are to be found in [20, Part IV, 2]. Hence the invariant, $\lambda(f)$, of the embedding f distinguishes cobordism types and is defined in a very general manner which uses only the transfer and the fact that (8.28) is invertible.

In the next section it will be shown how a $\lambda(f)$-invariant may be defined for a smooth algebraic embedding defined over an algebraically closed field of finite characteristic.

8.31 Problem. Give a direct *geometrical* version of the Pontrjagin-Thom construction in terms of AU-theory.

§9. Étale algebraic cobordism and smooth algebraic embeddings

9.1. Let K be an algebraically closed field of characteristic $p < \infty$. Several authors (we will follow [13]) have constructed an étale homotopy classifying space for GL_nK $(1 \leq n \leq \infty)$. In [12;13] this is called the restricted etale homotopy type for the classifying space of GL_nK and is written $W(GL_nK)_{ret}$. One may profinitely complete this pro-space away from p and form its associated inverse limit space [5] $BGL_nK^{\wedge}_{et}$. The union

$$BGLK^{\wedge}_{et} = \varinjlim_{n} BGL_nK^{\wedge}_{et}$$

is a simply connected H-space with π_2 equal to $\hat{Z}$ (finite completion of Z away from p). The results of [12;13;20, Part IV, §1] imply that $BGLK^{\wedge}_{et}$ is homotopy equivalent to $BU\hat{Z}$. Hence we obtain the following result.

9.2 Theorem. Let $\beta \in \pi_2(BGLK^{\wedge}_{et})$ be a generator (as a $\hat{Z}$-module) in the notation of §9.1. Let AK^*_{et} denote the étale cobordism theory associated with the spectrum $BGLK^{\wedge}_{et}(\beta)$.

Then

(a) $AK^0_{et}(\_) \cong MUZ^{2*}(\_)$ on the category of finite dimensional complexes.

(b) The Frobenius automorphism of K defined by $\phi_p(a) = a^p$ $(a \in K)$ induces a ring endomorphism of AK^*_{et} which corresponds under (a) to ψ^{-p}, the Adams operation of §8.10.

9.3. Now suppose that $\pi: E \to X$ is an algebraic vector bundle over a smooth K-variety, X. Both X and $E-X$ have prime-to-p profinite etale homotopy types and the map between them

$$\pi: (E-X)^{\wedge}_{et} \to X^{\wedge}_{et} \tag{9.4}$$

is the analogue of the induced spherical fibration of a topological vector bundle. We can form the cofibre of (9.4) to obtain a Thom space (a pro-space) $Th(E)^{\wedge}_{et}$. The étale cohomology of this pro-space is obtained by taking the direct limit of the cohomology of the spaces in the directed system. Similarly we may form the étale cobordism groups of X_{et} by applying AK^*_{et} and then taking a direct limit. Denote this by $\varinjlim AK^*_{et}(X^\wedge_{et})$.

9.5 Theorem. Let n be prime to p and set $hK^q(\_) = AK^q_{et}(\_;Z/n)$. Then, in the notation of §9.3, there is a Thom class

$$\Lambda(E)^\wedge_{et} \in \varinjlim hK^0(Th(E)^\wedge_{et})$$

which induces, by multiplication, a Thom isomorphism

$$\varinjlim hK^q(X^\wedge_{et}) \xrightarrow{\cong} \varinjlim hK^q(Th(E)^\wedge_{et}).$$

Also $\Lambda(E)_{et}$ may be chosen coherently as n varies.

Proof: By the standard spectral sequence arguments, the Thom class having been obtained by Theorem 9.2.

9.6. Suppose now that $f: X \to Y$ is a smooth algebraic embedding. We may form the "étale bordism" groups

$$(\varprojlim AK_{et})_*(X^\wedge_{et})$$

by taking the inverse limit of AK_{et}-homology groups. We may also form

$$(AK_{et})_*(\varprojlim X^\wedge_{et}).$$

We can attempt to define, after the method of §8.27, homomorphisms

$$\lambda(f): hK_q(\varprojlim X^\wedge_{et}) \to \pi_q(hK)$$

(9.7) and

$$\tilde{\lambda}(f): \varprojlim hK_q(X^\wedge_{et}) \to \pi_q(hK)$$

where hK_* is AK_{et}-homology. This will be possible if the analogues of §8.28 are isomorphisms

$$\varprojlim hK_q(Th(\nu)^\wedge_{et}) \to \varprojlim hK_q(X^\wedge_{et}) \tag{9.8}$$

and

$$hK_q(\varinjlim Th(\nu)^\wedge_{et}) \to hK_q(\varprojlim X^\wedge_{et}). \tag{9.9}$$

n §§9.8-9.9 ν is the normal bundle of f and hK_* is AK_{et}-homology. If (9.8) s an isomorphism we call f $\underleftarrow{\lim}\, AK_{et}$-*orientable* and $\tilde{\lambda}(f)$ exists. If (9.9) is n isomorphism we call f $AK_{et}\, \underleftarrow{\lim}$-*orientable* and $\lambda(f)$ exists. I am told that the hom isomorphism of §9.5 will suffice to define a $\lambda(f)$-type homomorphism if one uses he Steenrod homology theory associated to AK^*_{et}, but I have so far made no attempt o verify this.

In conclusion I must say that these orientability notions are not at all undertood and may even be the wrong ones. In general one would like to get away with equiring only §9.5 to hold. It is a consequence of a result of Artin-Mazur-Sullian that over the complex field $\lambda(f)$ exists under very mild restrictions (namely hat the set of complex points of X, in the strong topology, is a connected maniold) and when $Y = C^M$, $\lambda(f)$ captures the $\hat{Z}$-adic information in the Pontrjaginhom construction. This result follows from the discussion in 8.27 and is elaborted upon in [20, Part IV].

.10 Problem. Give an algebraic-geometric description of $\lambda(f)$ for a C-variety, say. erhaps, for this purpose, it is possible to define a pro-transfer in the geometrical etting before passing to etale homotopy types, thereby allowing a description simiar to that of §8.25.

.11 Problem. Give criteria for an algebraic embedding to be orientable in the enses mentioned in §9.6.

Follow up the Steenrod AK_{et}-homology approach and show that Theorem 9.5 implies he appropriate orientability for this approach.

REFRENCES

[1] J.F. Adams: Stable homotopy and generalized homology; Chicago Lecture Notes in Maths. (1974).

[2] J.F. Adams: Lectures on generalised cohomology; Lecture Notes in Maths. 99, Springer-Verlag (1969) 1-138.

[3] J.C. Becker and D.H. Gottlieb: The transfer map and fibre bundles; Topology 14 (1975) 1-12.

[4] J.C. Becker and D.H. Gottlieb: Transfer maps for fibrations and duality; Compositio Math (1977).

[5] A.K. Bousfield and D. Kan: Homotopy limits, Completions and localisations; Lecture Notes in Maths. 304, Springer-Verlag (1972).

[6] G. Brumfiel and I. Madsen: Evaluation of the transfer and the universal surgery classes; Inventiones Math. 32 (1976) 133-169.

[7] F. Cohen and L. Taylor: A stable decomposition for certain spaces; preprint (1977).

[8] P.E. Conner and E.E. Floyd: The relation of cobordism to K-theories; Lecture Notes in Maths. 28, Springer-Verlag (1966).

[9] A. Dold and R. Thom: Quasifaserungen und unendliche symmetriche produkte; Annals of Maths. (2) 67 (1958) 239-281.

[10] M. Feshbach: The transfer and compact Lie groups; Thesis, Stanford University (1976).

[11] Z. Fiedorowicz and S.B. Priddy: Homology of classical groups over finite fields and their associated infinite loopspaces; Northwestern University preprint (1977).

[12] E.M. Friedlander: Computations of K-theories of finite fields; Topology 15 (1976) 87-109.

[13] E.M. Friedlander: Exceptional isogonies and the classifying spaces of simple Lie groups; Annals of Maths 101 (1975) 510-520.

[14] A. Hattori: Integral characteristic numbers for weakly almost complex manifolds; Topology 5 (1966) 259-280.

[15] D.S. Kahn and S.B. Priddy: Applications of the transfer to stable homotopy theory; Bull. A.M. Soc. 741 (1972) 981-987.

[16] R. Lee and R.H. Szczarba: The group $K_3(Z)$ is cyclic of order forty-eight; Annals of Maths. 104 (1976) 31-60.

[17] I. Madsen, V.P. Snaith and J. Tornehave: Infinite loop maps in geometric topology; Math. Proc. Cambs. Phil Soc. (1977) 81, 399-430.

[18] R.E. Mosher: Some stable homotopy of complex projective space; Topology 7 (1968) 179-193.

[19] C.L. Reedy: Thesis, University of California at La Jolla (1975).

[20] V.P. Snaith: Algebraic cobordism and K-theory; to appear Mem. A.M. Soc.

[21] V.P. Snaith: On the S-type of imJ; Proc. Conf. on geometric topology and homotopy theory (Evanston, 1977), to appear in Springer-Verlag Lecture Notes in Maths.

[22] V.P. Snaith: On the algebraic cobordism of Z, submitted to Topology.

[23] V.P. Snaith: Stable decomposition of $\Omega^n\Sigma^n X$; J. London Math. Soc. 2 (7) (1974) 577-583.

[24] R.E. Stong: Relations among characteristic numbers, I; Topology 4 (1965) 267-281.

25] R. Thom: Quelques proprietes globales des varietes differentiables; Comm. Math. Helv. 28 (1954) 17-86.

26] J.B. Wagoner: Delooping the classifying spaces of algebraic K-theory; Topology 11 (1972) 349-370.

27] G. Wolff: Von Conner-Floyd theorem zum Hattori-Stong theorem; Manuscripta Math. 17 (1975) 327-332.

28] Problem Session, A.M. Soc. Summer Institute (1976) Stanford.

FIBRE PRESERVING MAPS AND FUNCTIONAL SPACES

Peter I. Booth, Philip R. Heath and Renzo A. Piccinini

0. Introduction.

Let $q: Y \to A$, $r: Z \to B$ be maps (= continuous functions). A fibre preserving map from q to r is a pair (g_1, g_0) of maps $g_1: Y \to Z$, $g_0: A \to B$ such that $rg_1 = g_0 q$. There is an obvious category whose objects are maps $q, r, \ldots$ and whose morphisms are fibre preserving maps; we denote these morphisms by $(g_1, g_0): q \twoheadrightarrow r$. (Please note there is no surjectivity connotation).

In practice $q, r, \ldots$ etc., will have a Covering Homotopy Property, i.e. be locally trivial or be Hurewicz fibrations. We are concerned with the construction and properties of a functional space $Y \cdot Z$, an associated fibration $q \cdot r: Y \cdot Z \to A \times B$ and with corresponding exponential laws; the set of cross-sections to the composite of $q \cdot r$ with the projection onto A, for example, is in one-to-one correspondence with the set of fibre preserving maps $q \twoheadrightarrow r$. We draw the reader's attention to the way these exponential laws differ from those in [3] and elsewhere, the main point being that our results generalize from being over a fixed base space to the situation of having variable base spaces.

After defining $q \cdot r$ and establishing its basic properties (section 1), we move directly on (in section 2) to our applications. By restricting a suitably chosen $q \cdot r$ to the subspace of homotopy equivalences in $Y \cdot Z$, we obtain a "Hurewicz fibration analogue" of Dold's Functional Bundle [8], solving a problem raised by Allaud [1, page 218]. Part of the purpose of this paper is to prepare the ground to discuss these considerations more extensively in [7]. We also show that the total space of Dold's Bundle is in fact a subspace of the corresponding $Y \cdot Z$. In a further application we solve a query of Maehara [10], showing that his obstruction theory for fibre preserving maps is simply the obstruction theory for cross-sections to $q \cdot r$.

The relationship between $q \cdot r$ and the first Author's fibred mapping spaces (c.f. [3], [4]) is discussed in section 3, showing that each is a special case of the other and thus enabling us to use his previous work as a basis for some of our

·oofs. The discussion in the main part of this paper takes place in the context ˙ the convenient category of k-spaces [2], i.e. the category HG of [11].

We append (in section 4) a brief introduction to the corresponding theory ·ing ordinary topological spaces.

Functional Exponential Laws and Fibre Preserving Maps.

We work in the convenient category of k-spaces [2], [11], i.e. spaces with ıe final topology with respect to all incoming maps from compact Hausdorff ·aces. Any space can be k-ified (retopologized as a k-space) by giving it the ·ove final topology. As is usual in a convenient category subspaces, products, ıllbacks, mapping spaces (Map(X,Y)) etc., are the k-ifications of the appropriate ›nstructions in Top. The appropriate topology for Map(X,Y) in Top is, of course, ıe compact-open topology.

If Z is a space we define a new space $Z^+ = Z \cup \{\infty\}$ to be the k-ification f the topology defined by requiring C to be closed in Z^+ if either $C = Z^+$ r if C is closed in Z. This construction enables us to identify partial maps $: X \to Z$, defined on a closed subset of X with continuous maps $\tilde{f}: X \to Z^+$ efined in the obvious way.

Given a T_1-space A and maps $q: Y \to A$, $r: Z \to B$, we define the set

$$Y \cdot Z = \bigcup_{a \in A, b \in B} \text{map}(Y_a, Z_b),$$

here Y_a, Z_b are the respective fibres (inverse images) of q and r over a nd b, and $\text{map}(Y_a, Z_b)$ is the set of maps from Y_a to Z_b. Define a function $: Y \cdot Z \to \text{Map}(Y, Z^+)$ by $j(f)(y) = f(y)$ if $y \in Y_a$, $f: Y_a \to Z_b$ and $j(f)(z) = \infty$, therwise. The condition that A is T_1 ensures that each $f \in Y \cdot Z$ has a losed domain when considered as a partial map from Y into Z. We now give $\cdot Z$ the (k-ified) initial topology with respect to j and $q \cdot r: Y \cdot Z \to A \times B$ efined by $q \cdot r(f: Y_a \to Z_b) = (a,b)$.

This technique is similar to that used in [5] to topologise (YZ). We will efer to $Y \cdot Z$ as a functional space; we also denote the composition of $q \cdot r$ ith the projection onto A as $q \cdot_1 r$.

We assume, from this point on, that A is k-Hausdorff, i.e. that the diagonal is closed in the k-ified product space $A \times A$. Hausdorff spaces are k-Hausdorff.

Theorem 1. The Functional Exponential Law - Let $p: X \to A$, $q: Y \to A$ and $r: Z \to B$ be maps. There is a one-to-one correspondence between (i) the set of fibre preserving maps $(f_1, f_0): q_p \twoheadrightarrow r$, and (ii) the set of maps $f: X \to Y \cdot Z$ over A, this correspondence being determined by $f_1(x,y) = f(x)(y)$.

Here q_p denotes the projection of the pullback space $X \sqcap Y$ to X. The result is illustrated by the diagram

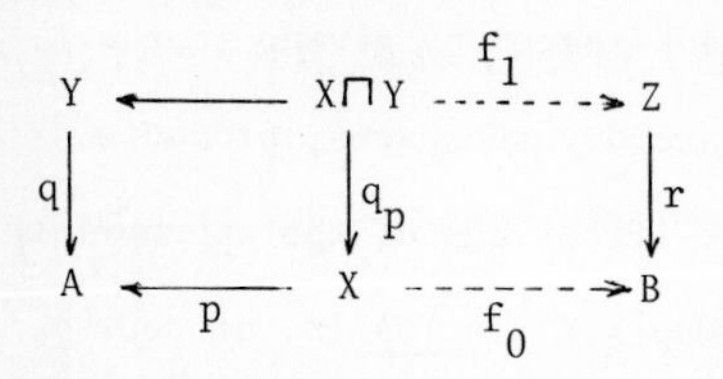

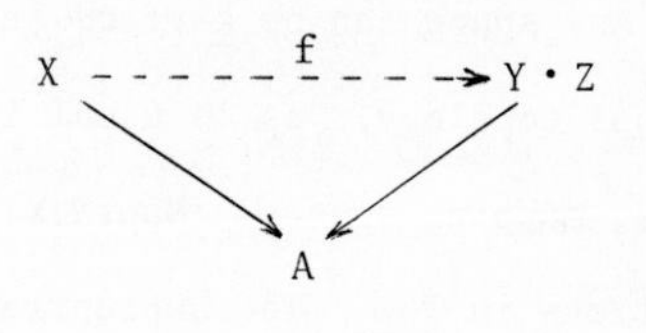

We notice that each f of (ii) determines the corresponding f_0 as the composite $(q \cdot_2 r)f$ where $q \cdot_2 r$ is the composite of $q \cdot r$ with the projection of $A \times B$ onto B. We notice also that if $A = B = *$ (a one point space) then the Theorem reduces to the ordinary exponential law of spaces.

Proof. The condition that A is k-Hausdorff ensures that $X \sqcap Y$ is closed in $X \times Y$ and hence maps $X \sqcap Y \to Z$ can be associated with maps $X \times Y \to Z^+$. The result now follows from properties of initial topologies and the ordinary exponential law. //

Taking p to be the identity on A we obtain the following result.

Corollary 2. There is a one-to-one correspondence between (i) the set of fibre preserving maps $(f_1, f_0): q \twoheadrightarrow r$, and (ii) the set of cross-sections to $q \cdot_1 r$.

Taking $p: A \times I \to A$ to be the projection onto A we obtain

Corollary 3. A pair of maps $(f_1, f_0): q \twoheadrightarrow r$ and (g_1, g_0) are homotopic via a family of fibre preserving maps if, and only if, the corresponding sections to $q \cdot_1 r$ are vertically homotopic.

Corollary 4. Given $a \in A$ the fibre of $q \cdot_1 r$ over a is the subspace of $\mathrm{Map}(Y_a, Z)$ consisting of maps whose images are contained in a single fibre of Z.

We will denote this space by $Z^{(Y_a)}$.

Proof. The identification of the underlying sets is clear; we merely observe that a function from a space into $Z^{(Y_a)}$ of some map is continuous if, and only if, the same function into $\mathrm{Map}(Y_a, Z)$ is continuous. //

Proposition 5. If $q: Y \to A$, $r: Z \to B$ are locally trivial maps then the topology on $Y \cdot Z$ agrees with the weak topology with respect to the local triviality structure that exists on $q \cdot r: Y \cdot Z \to A \times B$.

Proof. As will be shown in section 3(a) we can identify $Y \cdot Z$ with the space $(Y \times B \quad A \times Z)$. The k-space analogue of [6, Theorem 2.1] referred to in [6, section 8] now gives the result. //

Proposition 6. If q and r are Hurewicz fibrations then so is $q \cdot r : Y \cdot Z \to A \times B$.

Proof. Consider the fibrations $p \times 1: Y \times B \to A \times B$ and $1 \times r: A \times Z \to A \times B$ over the same base $A \times B$; then $(p \times 1 \quad 1 \times r): (Y \times B \quad A \times Z) \to A \times B$ is a fibration by [2, Theorem 3.4]. The result then follows from the identification of $q \cdot r$ with $(p \times 1 \quad 1 \times r)$ as outlined in section 3. //

Our next two results will not be used until the sequel [7]; however, the methods of proof are in keeping with the philosophy of this section.

Corollary 7. If q and r are Hurewicz fibrations and their fibres have the same homotopy type, then the restriction of $q \cdot r$ to the subspaces of $Y \cdot Z$ consisting of homotopy equivalences, is also a Hurewicz fibration.

Proof. It is sufficient to show that if $f: Y_a \to Z_b$ is a map in a path component of $Y \cdot Z$ containing a homotopy equivalence, then f is itself a homotopy equivalence. To this end let $\lambda: I \to Y \cdot Z$ be a path in $Y \cdot Z$ in which $\lambda(0): Y_a \to Z_b$ say, is a homotopy equivalence, $\lambda(1) = f$ and let $\mu = (q \cdot r)\lambda$. Then λ considered as a map from μ to $q \cdot r$ over $A \times B$ corresponds by Theorem 1 to a map pair represented by $(\theta, \mathrm{proj}_B \lambda)$ in the diagram

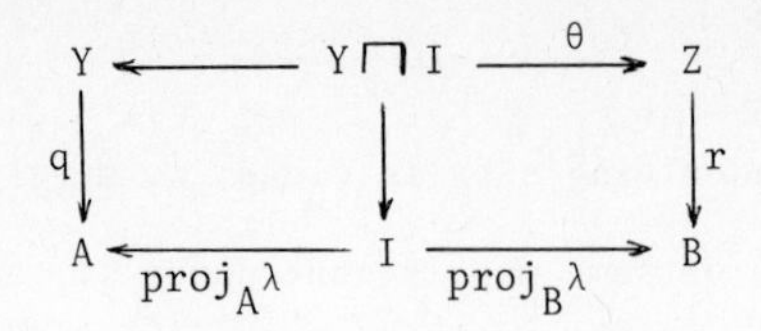

Now θ and $\mathrm{proj}_B\lambda$ determine a map $\phi\colon Y \sqcap I \to Z \sqcap I$ between fibrations over I. The restriction of ϕ to the fibre over 0 is easily seen to be $\lambda(0)$ which is of course a homotopy equivalence; it follows that the restriction to any fibre is a homotopy equivalence, and in particular, the restriction of ϕ to the fibre over 1. This restriction is seen to be $f = \lambda(1)$ giving the required result. //

Theorem 8. Let $f, g\colon A \to B$ and $p\colon E \to B$ be maps. Then there is a one-to-one correspondence between: (i) the lifts of $(f,g)\colon A \to B \times B$ over $p \bullet p\colon E \bullet E \to B \times B$, and (ii) the fibre preserving maps from $p_f\colon A \sqcap_f E \to A$ into $p_g\colon A \sqcap_g E \to A$ over A.

We remark that in the case in which a lift of (f,g) over $p \bullet p$ lands in the subspace of $E \bullet E$ consisting of homotopy equivalences (which is the case if f is homotopic to g) then by [8] the corresponding fibre map is a fibre homotopy equivalence. We shall use these ideas in [7].

Proof. Using the identifications as in the proofs of Propositions 5 and 6 (see also section 3) and the Fibred Exponential Law [3] over $B \times B$ we deduce the existence of a one-to-one correspondence between maps $k''\colon (f,g) \to p \bullet p$ over $B \times B$ and maps $k'\colon (f,g) \sqcap (p \times 1) \to 1 \times p$ over $B \times B$. We identify the total space of the pullback $(f,g) \sqcap (p \times 1)$ to $A \sqcap_f E$ via the double pullback diagram

$$\begin{array}{ccccc}
A \sqcap_f E & \longrightarrow & E \times B & \longrightarrow & E \\
\downarrow{\scriptstyle p_f} & & \downarrow{\scriptstyle p \times 1} & & \downarrow{\scriptstyle p} \\
A & \xrightarrow[(f,g)]{} & B \times B & \xrightarrow[\mathrm{proj}_1]{} & B
\end{array}$$

Similarly the pullback of (f,g) and $1 \times p$ is $A \sqcap_g E$ and so there is, using the universal property of pullbacks, a one-to-one correspondence between maps $k'\colon A \sqcap_f E \to B \times E$ over $B \times B$ and fibre maps $k\colon A \sqcap_f E \to A \sqcap_g E$ over A, as required. The argument is clearly reversible. //

2. The relations of $q \cdot r$ to Dold's construction and questions of Allaud and Maehara.

(i) Let $q: Y \to A$, $r: Z \to B$ be locally trivial principal G-bundles. It follows by a slight modification of Proposition 5, that the (k-ified version) of the functional bundle (q,r) of [8, page 249] is just the restriction of $q \cdot r$ to the appropriate subspace of $Y \cdot Z$. The key property of $(q,r): (Y,Z) \to A$, as might be expected, is that its cross-sections are in bijective correspondence with the fibre preserving G-maps $q \twoheadrightarrow r$.

(ii) In [1, page 218] G. Allaud discusses the possibility of producing a construction analogous to that of Dold, for Hurewicz fibrations $q: Y \to A$ and $r: Z \to B$. Assuming that the fibres of q and r are all of the same homotopy type and $H(Y_a, Z_b)$ denotes the set of homotopy equivalences of Y_a to Z_b, then the solution is to topologize the set

$$\bigsqcup_{a \in A, b \in B} H(Y_a, Z_b)$$

as a subspace of our $Y \cdot Z$. It then follows that the set of fibre preserving maps $q \twoheadrightarrow r$, whose restrictions to fibres are homotopy equivalences, is in bijective correspondence with the set of cross-sections to the projection of $\bigsqcup H(Y_a, Z_b)$ into A. This result is useful in the case where r is a Universal Hurewicz fibration and is taken up again in [7].

(iii) Given that (A,L) is a relative CW-complex and that $p: X \to A$, $q: Y \to B$ are Hurewicz fibrations, we denote $p^{-1}(L)$ by $X|L$ and let $p|L$ be the restriction $p|(X|L) : X|L \to L$. If $(g_1, g_0): p|L \twoheadrightarrow q$ is a fibre preserving map we consider the problem of setting up an obstruction theory for studying the existence and homotopy classification of fibre preserving extensions (f_1, f_0) of (g_1, g_0) over p, in other words, of completing the following diagram

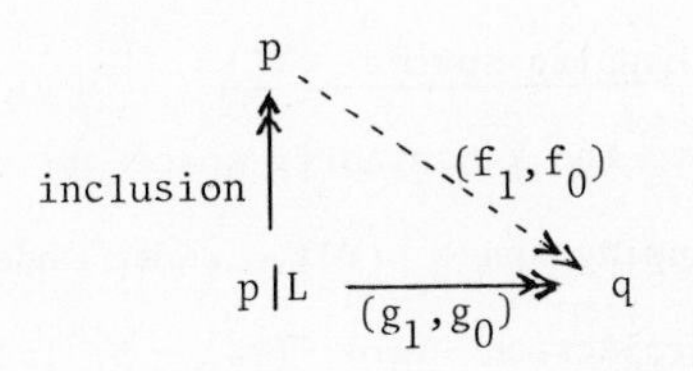

The relative n-skeleton $A^n \cup L$ of (A,L) will be denoted by $\overline{A}^n$; we define $\overline{X}^n = p^{-1}(\overline{A}^n)$ and $\overline{p}^n = p|\overline{X}^n\colon \overline{X}^n \to \overline{A}^n$. We assume that p is orientable and has a distinguished fiber F; let $\mu\colon F \to Y$ be the restriction of g_1.

<u>Theorem 9</u>. (Maehara) [10, page 60] - Given that $(h_1,h_0)\colon \overline{p}^n \twoheadrightarrow q$, the obstruction to extending its restriction to $\overline{p}^{n-1}$ over $\overline{p}^{n+1}$ is located in $H^{n+1}(A,L;$ $\pi_n(Y^{(F)},u))$.

<u>Theorem 10</u>. (Maehara) [10, page 64] - Given that the restrictions of $(h_1,h_0)\colon$ $p \twoheadrightarrow q$ and $(h_1',h_0')\colon p \twoheadrightarrow q$ to $\overline{p}^{n-1}$ are "fibre preserving homotopic", then the obstruction to extending the restriction of this homotopy to $\overline{p}^{n-2}$ into a fibre preserving homotopy between the restrictions of (h_1,h_0) and (h_1',h_0') to $\overline{p}^n$ is located in $H^n(A,L;\ \pi_n(Y^{(F)},u))$.

Several applications of these two theorems are given in [10, pages 72-111]; we will not however, be concerned with them here.

Maehara raises the possibility that the proofs of these theorems can be reduced to cross-section problems for an appropriate fibration [10, page 1]. It is clear that the properties of fibre preserving maps discussed in Theorems 9 and 10 reduce, via our Corollaries 2 and 3, to the existence and homotopy classification problems for cross-sections to $p \cdot q$ that extend a given partial section over L.

Our definition of <u>orientability</u> is slightly stronger than the usual one, for instead of requiring that loops in A induce homomorphisms of homology $H_*(F) \to H_*(F)$, we require that the induced maps $F \to F$ shall be homotopic to the identity. It can be shown (via Theorem 1) that p is orientable implies that $p \cdot q$ is orientable (this is immediate in the case where A is simply connected). Standard obstruction theory arguments for cross-sections applied to the fibration $p \cdot q$ now reprove Theorems 9 and 10 [9].

3. Functional spaces $Y \cdot Z$ and fibred mapping spaces (YZ)

Let $q\colon Y \to B$, $r\colon Z \to B$ be maps into the <u>k</u>-Hausdorff space B. We remind the reader of the existence of the fibred mapping space (YZ), whose underlying set is $\bigcup_{b\in B} \mathrm{Map}(Y_b,Z_b)$, and of the obvious projection $(qr)\colon (YZ) \to B$ (as defined in

2], [5]). The question arises as to which of the mapping spaces (YZ) or $Y \cdot Z$ s "most basic". We show below that each can be defined in terms of the other, o neither can be regarded as being more fundamental.

We assume, for the remainder of this section, that A and B are k-Hausdorff.

(a) Given maps $q: Y \to A$ and $r: Z \to B$ define the composite map

$$(Y \times B \quad A \times Z) \xrightarrow[(q \times 1_B \quad 1_A \times r)]{} A \times B \xrightarrow[\text{proj}]{} A$$

nd notice that the underlying set of $(Y \times B \quad A \times Z)$ can be identified with hat of $Y \cdot Z$. We apply the exponential law for (qr) [3] to show that the revious composite satisfies Theorem 1, and hence can be identified with $q \cdot r$.

(b) Given $q: Y \to B$, $r: Z \to B$ we define $Y \cdot Z$ and the projection $q \cdot r: Y \cdot Z \to B \times B$. Now the diagonal $\Delta: B \to B \times B$ induces by pullback a projection

$$(q \cdot r)_{\Delta}: (Y \cdot Z) \sqcap B \to B$$

whose underlying function can be identified with $(qr): (YZ) \to B$. Our exponential law for $Y \cdot Z$ (Theorem 1) ensures that maps into $(Y \cdot Z) \sqcap B$ satisfy the same exponential law [3] as do maps into (YZ), hence $(qr): (YZ) \to B$ can be defined using the above projection.

4. Functional spaces in the category of all topological spaces

The results of the previous sections can also be obtained in the category of all topological spaces, subject to some restrictions on the spaces involved.

We define Z^+ as the non-k-ified Z^+ of section 1. If $q: Y \to A$, $r: Z \to B$ are maps and A is a T_1-space, we give the set $Y \cdot Z$ the initial topology with respect to the functions

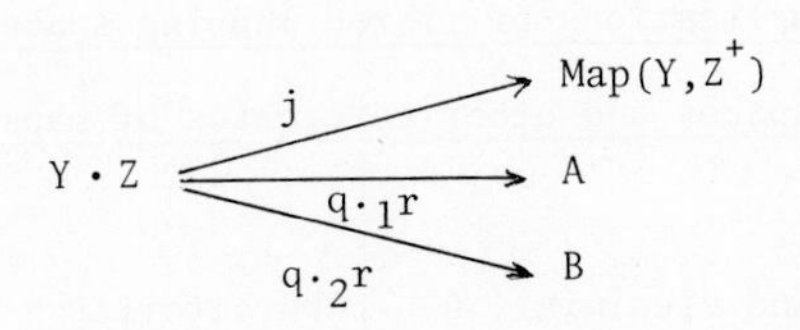

where $\mathrm{Map}(Y,Z^+)$ is given the compact-open topology. Hence $Y \cdot Z$ has a subbasis consisting of three types of open sets: (a) $W(K,U) = \{f \in Y \cdot Z \mid f(K) \subset U\}$, where K is compact in Y, U is open in Z and $f(K) = \{f(x) \mid x \in K \cap \text{dom } f\}$; (b) $(q \cdot_1 r)^{-1}(V)$, where V is open in A; (c) $(q \cdot_2 r)^{-1}(W)$, where W is open in B.

Various conditions, analogous to those appearing in [5], are sufficient for the following exponential law; we give the most obvious case.

Theorem 11. Functional Exponential Law - Let us assume that $p: X \to A$, $q: Y \to A$, $r: Z \to B$ are maps, where A is a Hausdorff space and Y is locally compact Hausdorff. Then there exist a bijective correspondence between:

(i) the set of fibre preserving maps $(f_1, f_0): q_p \twoheadrightarrow r$, and

(ii) the set of maps $f: X \to Y \cdot Z$ over A,

determined by $f_1(x,y) = f(x)(y)$, $p(x) = q(y)$.

The proof of this theorem, and of analogues of the other results above, can be obtained by mimicking previous arguments in the context of the category Top.

References

[1] Allaud, G. - Concerning universal fibrations and a theorem of E. Fadell, Duke Math. J. 37, 213-224 (1970).

[2] Booth, P. - The section problem and the lifting problem, Math. Z. 121, 273-287 (1971).

[3] Booth, P. - The exponential law of maps II, Math. Z. 121, 311-319 (1971).

[4] Booth, P. - A unified treatment of some basic problems in homotopy theory, Bull. Amer. Math. Soc. 79, 331-336 (1973).

[5] Booth, P. and Brown, R. - Spaces of partial maps, fibred mapping spaces and the compact-open topology. To appear in Gen. Top. and its applics.

[6] ________ - On the applications of fibred mapping spaces to exponential laws for bundles, ex-spaces and other categories of maps. To appear in Gen. Top. and its applics.

[7] Booth, P., Heath, P. and Piccinini, R. - Characterizing Universal Fibrations These Proceedings.

[8] Dold, A. - Partitions of Unity in the theory of fibrations. Ann. of Math. 78 (2), 223-255 (1963).

[9] Inoue, Y. - On singular cross-sections. Proc. Japan Acad. 31, 678-681 (1955).

[10] Maehara, R. - An obstruction theory for fibre preserving maps, Ph.D. thesis, Iowa State University (1972).

[11] Vogt, R. - Convenient categories of topological spaces for homotopy theory, Arch. Math. 22, 546-555 (1971).

CHARACTERIZING UNIVERSAL FIBRATIONS

Peter I. Booth, Philip R. Heath and Renzo A. Piccinini

Introduction - For the purposes of this introduction, we use the term "Fibration" to denote Hurewicz fibration, Principal G-bundle, Principal H-fibration or some similar notion. The concept of Universal fibration as discussed in the literature assumes various forms that can be defined within the context of a single class of fibrations. We distinguish four types. A fibration $p_\infty : E_\infty \to B_\infty$ is said to be (i) Free Universal if the appropriately defined equivalence classes of fibrations over a space B are classified by the free homotopy classes $[B,B_\infty]$; we say that p_∞ is (ii) Grounded Universal if the analogous grounded equivalence classes of fibrations over B are classified by $[B, B_\infty]_*$; we say that p_∞ is (iii) Aspherical Universal if the total space of the associated principal fibration is weakly contractible (i.e., aspherical) and finally, (iv) p_∞ is Extension Universal (c.f. [14], 19.2) if any partial map pair into p_∞ can be extended.

Various connections between the above have been exhibited in the literature. Dold in [8] has shown the equivalence of (i) and a strengthened form of (iii) for Principal G-bundles, while Steenrod in [14] has shown (iv) implies (i) in the same context. For Hurewicz fibrations, Allaud in [1] shows that if the fibres have the homotopy type of a CW-complex then (ii) implies (iii) and in [2] that the contractibility of the total space of the associated principal fibration implies (ii).

The aim of this paper is the systematic study of these connections for the various classes of fibrations and in particular, to determine when the term "Universal Fibration" carries no ambiguity. We use a general framework similar to that used by J. P. May in [11] to discuss the existence of Free Universal fibrations. In this context, we show (section 3) that (iii) and (iv) are equivalent and that these two imply (i) and (ii). We also observe that (ii) implies (i) (the direct proof (iii) implies (i) - Theorem 3.2 - is not redundant, because of its relation to comment (3) in section 5). In the examples section (section 4), we show the equivalence of all four types of universality for each of the specific examples mentioned at the beginning of this Introduction; also, we give an example of a fibration that is Universal in the senses (i) and (ii) but not (iii) or (iv). Relaxing the conditions in our category of fibrations we exhibit an example that is Universal in sense (ii) but in none of the other senses. The connections between the now unambiguous Universal fibrations, in the differing interpretations of the word fibration, are discussed (section 5).

The technique in our proofs is to show that each type of universality corresponds to properties of an appropriate class of functional fibrations. Given that p and p_∞ are "fibrations" (elements of our admissible category $\mathcal{A}$) then there is a functional fibration $p*_1p_\infty$ defined as a restriction of $p \cdot_1 p_\infty$ of [7]. If we fix p_∞ and allow p to range over the class of fibrations in question we see: (i) p_∞ is Free Universal is equivalent to (a weakened version of) the statement that each of the $p*p_\infty$ has a unique vertical homotopy class of sections; (ii) p_∞ is Grounded Universal is equivalent to the analogous statement for (i) with vertical based homotopy [5] replacing vertical homotopy; (iii) p_∞ is Aspherical Universal asserts that the $p*p_\infty$ have aspherical fibres and finally, (iv) p_∞ is Extension Universal is equivalent to each $p*p_\infty$ being a weak homotopy equivalence.

The paper is divided into five sections; the first discusses the foundations of our theory, the second makes formal our definitions and ensures the existence of certain required functors (in order to define equivalence in the definitions). The contents of sections 3 to 5 have already been discussed.

1 - <u>Foundations</u> - In what follows we shall work in the context of the convenient category $\mathcal{K}$ of <u>k</u>-spaces as in [7].

We begin by borrowing some notation and terminology from J.P. May [11]. Let $\mathcal{F}$ be a category with a distinguished object F together with a faithful underlying space functor $\mathcal{F} \to \mathcal{K}$. Thus each object of $\mathcal{F}$ is a <u>k</u>-space and the set $\mathcal{F}(X,Y)$ of morphisms from X to Y in $\mathcal{F}$ is a subset of $\mathcal{K}(X,Y)$; for technical reasons we shall assume that $\mathcal{F}(F,X) \neq \emptyset$, for every object X of $\mathcal{F}$. An <u>$\mathcal{F}$-space</u> is a morphism $p:X \to A$ of $\mathcal{K}$ such that A is a CW-complex and, for every $a \in A$,

$p^{-1}(a) \in \operatorname{Obj}\mathcal{F}$. Notice that if $*$ is a one-point space, the constant map $F \to *$ is an $\mathcal{F}$-space. An $\mathcal{F}$-map $(f_1, f_0): p \to r$ is a commutative diagram of $\mathcal{K}$

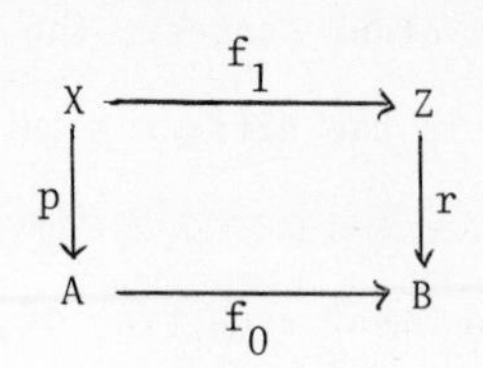

such that $f_1: p^{-1}(a) \to r^{-1}(f_0(a))$ is a morphism of $\mathcal{F}$, for all $a \in A$. If $A = B$ and $f_0 = 1_A$, f_1 is said to be an $\mathcal{F}$-map over B. An $\mathcal{F}$-homotopy is an $\mathcal{F}$-map (H,h) of the form

$$\begin{array}{ccc} X \times I & \xrightarrow{H} & Z \\ {\scriptstyle p \times 1_I}\downarrow & & \downarrow{\scriptstyle r} \\ A \times I & \xrightarrow[h]{} & B \end{array}$$

(we agree to identify $X \times *$ and $* \times X$ with X, for every object X of $\mathcal{F}$). In the case where $A = B$ and h is the projection, we have the notion of $\mathcal{F}$-homotopy over B. An $\mathcal{F}$-map $g: X \to Z$ over B is an $\mathcal{F}$-homotopy equivalence if there is an $\mathcal{F}$-map $g': Z \to X$ over B such that gg' and $g'g$ are $\mathcal{F}$-homotopic over B to the respective identity maps.

From now on we shall assume that $\mathcal{F}$ satisfies also the following condition:

(1.1) every morphism of $\mathcal{F}$ is an $\mathcal{F}$-homotopy equivalence over a point.

In what follows we shall be concerned mainly with a certain non-empty, full subcategory $\mathcal{A}$ - which we call admissible - of the category of $\mathcal{F}$-spaces and $\mathcal{F}$-maps. The objects of $\mathcal{A}$ will be called $\mathcal{A}$-fibrations; this name is suggested by J.P. May's work ([11], Def. 2.1) and Proposition 1.4. Before giving the axioms which define $\mathcal{A}$ we make a few remarks. Given an $\mathcal{F}$-space $r: Z \to B$ and a map $f: A \to B$ we denote the pull-back space $\{(a,z) \in A \times Z \mid f(a) = r(z)\}$ by $A \sqcap_f Z$. According to ([11], Lemma 1.2), the obvious map $r_f: A \sqcap_f Z \to A$ is an $\mathcal{F}$-space. Given maps $q: Y \to A$ and $r: Z \to B$, we denote by $q*r: Y*Z \to A \times B$ the restriction of $q \cdot r: Y \cdot Z \to A \times B$ (see [7]) to the subspace $Y * Z$ of $\mathcal{F}$-homotopy equivalences in $Y \cdot Z$; observe that the underlying set of $Y * Z$ is $\bigcup_{a\in A, b\in B} \mathcal{F}(Y_a, Z_b)$.

We are now ready to give the axioms for $\mathcal{A}$.

A1 - $F \to *$ is an $\mathcal{A}$-fibration.

A2 - If $r: Z \to B$ is an $\mathcal{A}$-fibration, A is a CW-complex and $f: A \to B$ is a map, then $r_f: A \sqcap_f Z \to A$ is an $\mathcal{A}$-fibration.

A3 - If $r: Z \to B$ is an $\mathcal{A}$-fibration, $s: W \to B$ is an $\mathcal{F}$-space and $g: Z \to W$ is an $\mathcal{F}$-map over B which is a homeomorphism, then s is an $\mathcal{A}$-fibration.

A4 - If $q: Y \to A$ and $r: Z \to B$ are $\mathcal{A}$-fibrations, then $q*r: Y * Z \to A \times B$ has the Covering Homotopy Property with respect to all CW-complexes.

Notice that A4 implies that the compositions of $q*r$ with the projections of $A \times B$ on the first or second factors also have the Covering Homotopy Property with respect to all CW-complexes; these compositions are, of course, the restrictions of $q \cdot_1 r$ and $q \cdot_2 r$ (see [7]) to $Y * Z$ and will be denoted by $q *_1 r$ and $q *_2 r$ respectively.

Theorem 1 and Corollaries 2 and 3 of [7] can be adapted to our present situation, yielding the following two Lemmas.

Lemma 1.2 - There is a one-to-one correspondence between the set of $\mathcal{F}$-maps $(f_1, f_0): q_p \to r$ and the set of maps $f: X \to Y*Z$ over A.

Lemma 1.3 - There is a one-to-one correspondence between the set of $\mathcal{F}$-maps from q to r and the set of cross-sections to $q *_1 r$. Furthermore, any such pair of $\mathcal{F}$-maps are $\mathcal{F}$-homotopic if, and only if, the corresponding cross-sections are vertically homotopic.

Proposition 1.4 - Let $q: Y \to A$ and $r: Z \to B$ be $\mathcal{A}$-fibrations and let (f_1, f_0): $q \to r$ be an $\mathcal{F}$-map. For every homotopy $h: A \times I \to B$ of f_0, there is a homotopy $H: Y \times I \to Z$ such that (H,h) is an $\mathcal{F}$-homotopy.

Proof - Construct the diagram

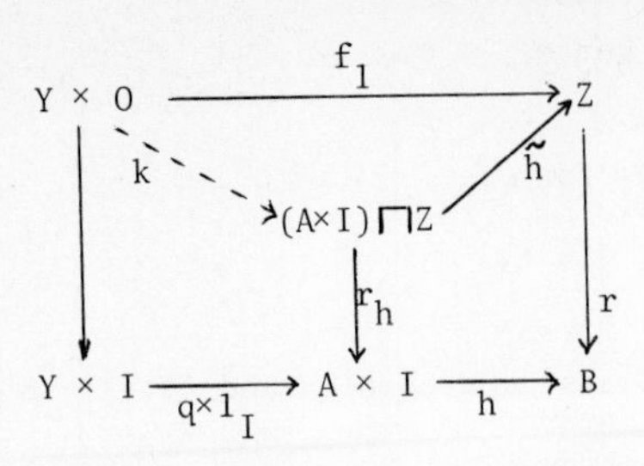

and observe that $r_h \in \mathrm{Obj}\mathcal{A}(\tilde{h},h)$ is an $\mathcal{F}$-map and that k is an $\mathcal{F}$-map over $A \times 0$. We shall show that there is an $\mathcal{F}$-map $H'\colon Y \times I \to (A\times I)\sqcap Z$ over $A \times I$ such that $H' \mid Y \times 0 = k$; then we just set $H = \tilde{h}\cdot H'$. To show the existence of H', we observe that since k is an $\mathcal{F}$-map there is a cross-section s of

$$(Y \times 0) * (A \times 0)\sqcap Z \xrightarrow{\quad (q \times 1)*_1 r_h \quad} A \times 0$$

by Lemma 1.3; composing s with the inclusion i of $(Y \times 0) * (A\times 0)\sqcap Z$ into $(Y \times I) * (A \times I)\sqcap Z$ we obtain a commutative diagram

$$\begin{array}{ccc} A \times 0 & \longrightarrow & (Y \times I) * (A \times I)\sqcap Z \\ \big\downarrow & & \big\downarrow (q\times 1) *_1 r_h \\ A \times I & \xrightarrow{\quad 1_{A\times I} \quad} & A \times I \end{array} .$$

Using A4, we obtain a cross-section K to $(q\times 1) *_1 r_h$ (whose restriction to $A \times 0$ is $i\circ s$). We apply Lemma 1.3 to K to obtain H'. //

2 - Universal $\mathcal{A}$-fibrations - We notice that Proposition 1.4 together with the obvious modifications of 2.4 and 2.5 of [11] give the following.

Proposition 2.1 - Let $p\colon E \to B$ be an $\mathcal{A}$-fibration and let f, g be homotopic maps from A into B, where A is a CW-complex. Then the $\mathcal{A}$-fibrations p_f and p_g are $\mathcal{F}$-homotopy equivalent over A. //

Given a CW-complex B, let $\mathcal{E}\mathcal{A}(B)$ - assumed to be a set - be the family of all $\mathcal{F}$-homotopy equivalence classes of $\mathcal{A}$-fibrations over B. In what follows, $\underline{HCW}$ and $\underline{HCW}_*$ will denote the respective homotopy categories of CW-complexes and based CW-complexes. The previous Proposition has the following consequence.

Corollary 2.2 - $\mathcal{E}\mathcal{A}$: HCW $\to$ Set is a contravariant functor.

One should also notice that the $\mathcal{A}$-fibration $p: E \to B$ defines a natural transformation

$$[\ ,B] \to \mathcal{E}\mathcal{A}(\ ),$$

where $[A,B]$ is the set of all homotopy classes of maps from A into B. This suggests the following.

Definition 2.3 - An $\mathcal{A}$-fibration $p_\infty: E_\infty \to B_\infty$ is said to be Free Universal amongst $\mathcal{A}$-fibrations if $[\ ,B_\infty] \to \mathcal{E}\mathcal{A}(\ )$ is a natural equivalence.

Remark 2.4 - If $p_\infty: E_\infty \to B_\infty$ is Free Universal, B_∞ is path-connected. This follows from the fact that if $b, b' \in B_\infty$ the inclusions

$$b, b': * \to B_\infty$$

induce $\mathcal{A}$-fibrations which are $\mathcal{F}$-homotopy equivalent to $F \to *$ over $*$, and so $b \simeq b'$.

Let us assume now that the CW-complex B has a base point $*$. A grounded $\mathcal{A}$-fibration $\xi = (p,k)$ is a sequence

$$F \xrightarrow{\ k\ } E_* = p^{-1}(*) \hookrightarrow E \xrightarrow{\ p\ } B$$

such that k is an $\mathcal{F}$-homotopy equivalence over $*$ and $p: E \to B$ is an $\mathcal{A}$-fibration. A morphism $(f_1,f_0): (p,k) \to (p',k')$ of grounded $\mathcal{A}$-fibrations is an $\mathcal{F}$-map $(f_1,f_0): p \to P'$ such that f_0 is based and $g|E_* \circ k$ is $\mathcal{F}$-homotopic to k'. If $B = B'$ and $f_0 = 1_B$ then by (1.1) and ([11], 2.6), f_1 is an $\mathcal{F}$-homotopy equivalence over B; we shall call such a morphism a grounded homotopy equivalence over B. It follows from (1.4) and ([11], 2.6) that this is an equivalence relation.

The following Lemma is a consequence of (1.1), ([11], 2.6) and Theorem 8 of [7].

Lemma 2.5 - Let $r: Z \to B$ be an $\mathcal{A}$-fibration and let f, g be maps from a CW-complex A into B. Then there is a one-to-one correspondence between: 1) lifts of $(f,g): A \to B \times B$ over $r * r$; 2) $\mathcal{F}$-homotopy equivalences of $r_f \to r_g$ over A. //

We shall now prove a result which we need for the notion of "grounded universality".

Theorem 2.6 - Let (p,k) be a grounded $\mathcal{A}$-fibration over B, (A,a_o) a based CW-complex and let $f, g : A \to B$ be base-homotopic. Then (p_f,k) and (p_g,k) ar grounded homotopy equivalent.

Proof - We know from the previous Lemma that corresponding to the identity $1: p_f \to p_f$ there is a lifting $\theta: A \to E * E$ of (f,f) over $p * p$. Consider the commutative diagram

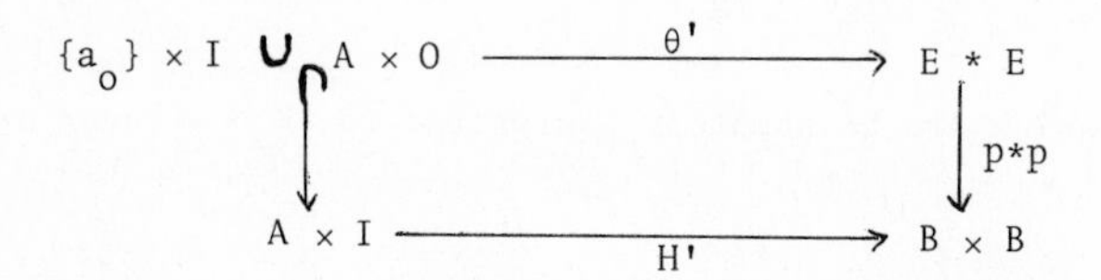

where H' is the based homotopy from (f,f) to (f,g) and θ' is such that $\theta'(a,t) = \theta'(a)$ for every $t \in I$ and $\theta' \mid A \times 0 = \theta$. Axiom A4 and Theorem 7.8.9 of [12] imply that there is a map $K: A \times I \to E * E$ completing the diagram. The restriction of K to $A \times 1$ corresponds by (2.5) to the required $\mathcal{F}$-homotopy equivalence of p_f into p_g. //

For a given based CW-complex A, let $\mathcal{E}\mathcal{A}^F(A)$ denote the family - assumed to be a set - of all grounded homotopy equivalence classes of grounded $\mathcal{A}$-fibrations over A.

Corollary 2.7 - $\mathcal{E}\mathcal{A}^F$: $\underline{HCW_*} \to \underline{Set_*}$ is a contravariant functor; furthermore, the grounded $\mathcal{A}$-fibration (p,k) defines a natural transformation $[\ ,B]_* \to \mathcal{E}\mathcal{A}^F(\ )$.

Definition 2.8 - Let (p_∞,k) be a grounded $\mathcal{A}$-fibration. Then (p_∞,k) is said to be Grounded Universal in $\mathcal{A}$ if $[\ ,B_\infty]_* \to \mathcal{E}\mathcal{A}^F(\ )$ is a natural equivalence.

If for all choices of $b \in B_\infty$ and all $\mathcal{F}$-maps $k: F \to E_{\infty b}$ the pair (p_∞,k) is Grounded Universal, then p_∞ is said to be Grounded Universal in $\mathcal{A}$.

Given any $\mathcal{A}$-fibration $p: E \to B$, we take the $\mathcal{A}$-fibration $c: F \to *$ and form $c * p: F * E \to * \times B$. Notice the the fibre of $c *_2 p$ over $b_o \in B$ is $F * F$, the k-space of all $\mathcal{F}$-homotopy equivalences of F into itself; by analogy with a standard construction in the theory of Hurewicz fibrations we call $c *_2 p = \mathrm{prin}_F p$ and $F * E = \mathrm{Prin}_F E$.

Definition 2.9 - An $\mathcal{A}$-fibration $p_\infty: E_\infty \to B_\infty$ is said to be Aspherical Universal if $\pi_n(\mathrm{Prin}_F E_\infty) = 0$, for all n and all choices of base point of $\mathrm{Prin}_F E_\infty$.

Definition 2.10 - An $\mathcal{A}$-fibration $p_\infty: E_\infty \to B_\infty$ is said to be Extension Universal in $\mathcal{A}$ if for every pair of CW-complexes (B,L) and every $\mathcal{A}$-fibration $p: E \to B$, each $\mathcal{F}$-map (f_{1L},f_{0L}) of the restriction $p|L$ of p to L onto p_∞ extends to an $\mathcal{F}$-map $(f_1,f_0): p \to p_\infty$.

3 - Relations between the various kinds of Universality - We begin this section by observing that every Grounded Universal $\mathcal{A}$-fibration is Free Universal. In fact, let (p_∞,k_∞) be Grounded Universal. For each CW-complex B let $B^+ = B \cup \{*\}$ and define a bijection $\mathcal{E}\mathcal{A}(B) \cong \mathcal{E}\mathcal{A}^F(B^+)$ by taking the class of an arbitrary $\mathcal{A}$-fibration over B into the class of $(p \cup c, 1_F)$, where $c: F \to *$($(p \cup c, 1_F)$ is a grounded $\mathcal{A}$-fibration by A2 and A3). Hence,

$$\mathcal{E}\mathcal{A}(B) \cong \mathcal{E}\mathcal{A}^F(B^+) \cong [B^+,B_\infty]_* \cong [B,B_\infty].$$

The reader should notice that Axioms A1 and A4 have not been used in the above observation.

Theorem 3.1 - An $\mathcal{A}$-fibration $p_\infty: E_\infty \to B_\infty$ is Aspherical Universal if, and only if, it is Extension Universal.

Proof - Necessity: Let (B,L) be a pair of CW-complexes, $p: E \to B$ be an $\mathcal{A}$-fibration and $(f_{1L},f_{0L}): p|L \to p_\infty$ be a given $\mathcal{F}$-map. Let s_L be the section of

$p|L *_1 p_\infty: E_L * E_\infty \to L$ which corresponds to (f_{1L}, f_{0L}) by (1.3). Then, if j and i are the inclusions $E_L * E_\infty \subset E * E_\infty$ and $L \subset B$, then $p *_1 p \cdot j \cdot s_L = 1_B \cdot i$. On the other hand, the Asphericity of p_∞ implies that $p *_1 p_\infty$ is a weak homotopy equivalence (the bijection $\pi_0(E * E_\infty) \cong \pi_0(B)$ follows by the exact homotopy sequence and the definitions). Hence, by ([12], 7.6.22) there is a map $s: B \to E * E_\infty$ such that $p *_1 p_\infty \cdot s = 1_B$ and $s \cdot i = j \cdot s_L$. The now familiar argument furnishes an $\mathcal{F}$-map $(f_1, f_0): p \to p_\infty$ which corresponds to s; furthermore, (f_1, f_0) extends (f_{1L}, f_{0L}).

<u>Sufficiency</u>. We wish to show that $\pi_n(\mathrm{Prin}_F E_\infty, k) = 0$, for all n and for an arbitrarily fixed base point k. Let $f, g: S^n \to \mathrm{Prin}_F E_\infty$ be base-preserving maps which correspond by (1.2) to $\mathcal{F}$-maps (f_1, f_0) and (g_1, g_0) from $S^n \times F \to S^n$ into p_∞. Consider the commutative diagram

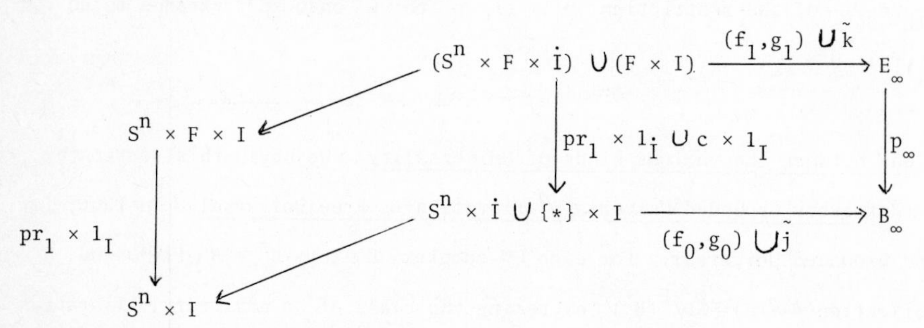

(where $\tilde{k}(y,t) = k(y)$, for every $t \in I$) and notice that since p_∞ is Extension Universal, there is an $\mathcal{F}$-map $(h_1, h_0): pr_1 \times 1_I \to p_\infty$ completing it. Again, applying (1.2) we obtain a base-homotopy $H: f \simeq g: S^n \times I \to \mathrm{Prin}_F E_\infty$ corresponding to (h_1, h_0). //

<u>Theorem 3.2</u> - Every Aspherical Universal $\mathcal{A}$-fibration is Free Universal.

<u>Proof</u> - Let $p_\infty: E_\infty \to B_\infty$ be an Aspherical Universal $\mathcal{A}$-fibration; we have to show that for every CW-complex B, $(p_\infty)_*(B): [B, B_\infty]_* \cong \mathcal{E}\mathcal{A}(B)$.

Let us first show that $(p_\infty)_*$ in onto. Let $p: E \to B$ be an $\mathcal{A}$-fibration; as in the previous Theorem, $p *_1 p_\infty$ is a weak homotopy-equivalence. Hence, by

([12], 7.6.22) applied to the CW-pair $(B, \emptyset)$ we obtain a section of $p *_1 p_\infty$ which in turn defines (uniquely) an $\mathcal{F}$-map $(f_1, f_0)\colon p \to p_\infty$. Consider the diagram

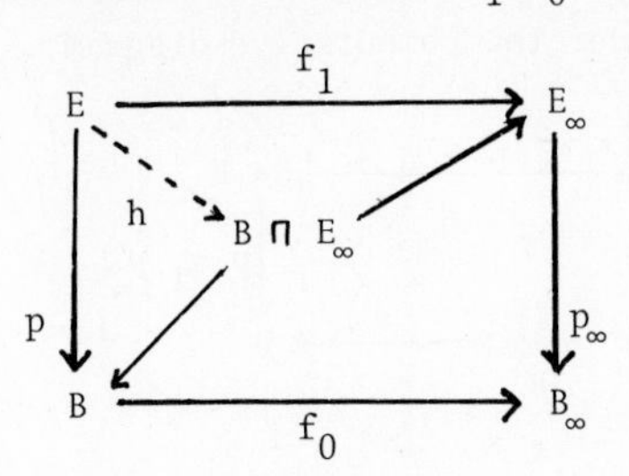

where h is the map obtained by the universal property of pull-backs. It follows from ([11], 1.2 and 2.6) and (1.1) that h is an $\mathcal{F}$-homotopy equivalence over B.

$(p_\infty)_*$ is one-to-one. Let $f, g\colon B \to B_\infty$ be such that $p_{\infty f}\colon B \sqcap_f E_\infty \to B$ and $p_{\infty g}\colon B \sqcap_g E_\infty \to B$ are $\mathcal{F}$-homotopy equivalent over B. The diagrams

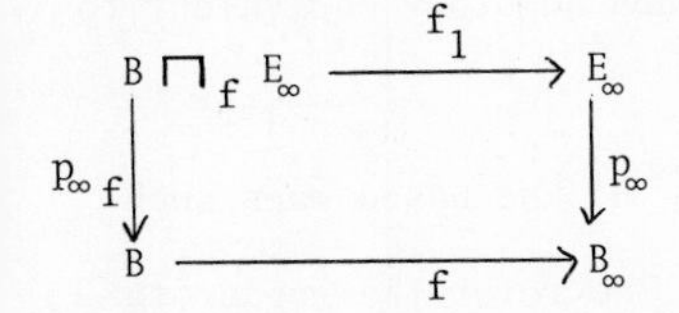

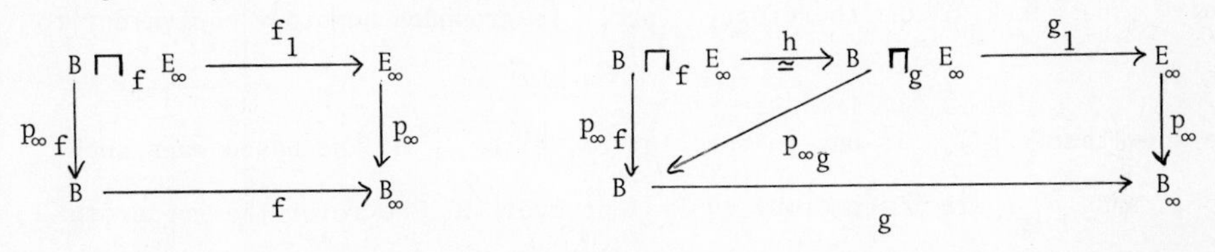

define two $\mathcal{F}$-maps (f_1, f) and $(g_1 h, g)$ from $p_{\infty f}$ into p_∞. These define two cross-sections $s(f)$, $s(g)$ of $p_{\infty f} *_1 p_\infty$. Consider the commutative diagram

$$\begin{array}{ccc} B \times \dot{I} & \xrightarrow{(s(f),s(g))} & (B \sqcap_f E_\infty) * E_\infty \\ \big\downarrow & & \big\downarrow{\scriptstyle p_{\infty f} *_1 p_\infty} \\ B \times I & \xrightarrow[pr_1]{} & B \end{array}$$

and notice that because of ([12], 7.6.22) there is a vertical homotopy $H\colon B \times I \to (B \sqcap_f E_\infty) * E_\infty$ of $s(f)$ into $s(g)$. It follows by (1.3) that (f_1, f) and $(g_1 h, g)$ are $\mathcal{F}$-homotopic. //

Theorem 3.3 - Every Aspherical Universal $\mathcal{A}$-fibration is Grounded Universal.

Proof - We want to show that for every based CW-complex B, $(p_\infty)_*\colon [B, B_\infty]_* \cong \mathcal{E}\mathcal{A}^F(B)$. We begin by showing that $(p_\infty)_*$ is onto. Suppose that $b \in B$ is the base point and let

$$F \xrightarrow{k} E_b \hookrightarrow E \xrightarrow{p} B$$

be a grounded $\mathcal{A}$-fibration. Consider the commutative diagram

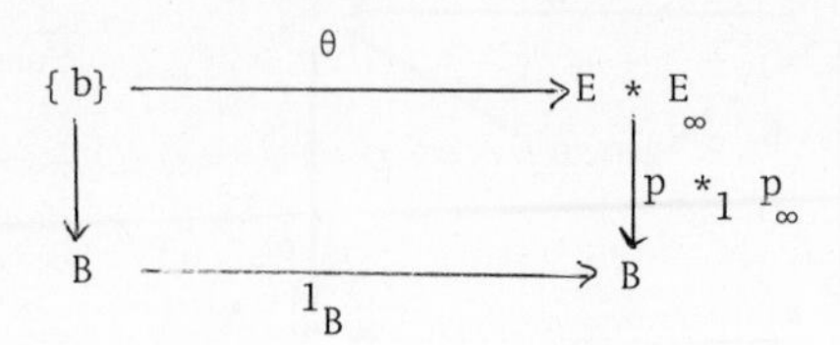

where $\theta(b) = k_\infty k^{-1}$; by ([12], 7.6.22) there is a cross-section s of $p *_1 p_\infty$ such that $s(b) = k_\infty k^{-1}$. It follows from Lemma 1.3 that s defines an $\mathcal{F}$-map $(f_1, f_0)\colon p \to p_\infty$ and as in (3.2), the induced map $h\colon E \to B \sqcap E_\infty$ is an $\mathcal{F}$-homotopy equivalence over B. On the other hand, the restriction of h to E_b is precisely $s(b) = k_\infty k^{-1}$ and therefore, (p,k) is grounded homotopy equivalent to $(p_{\infty f_0}, k_\infty)$.

To show that $(p_\infty)_*$ is one-to-one, let $f, g\colon B \to B_\infty$ be based maps such that $p_{\infty f}$ and $p_{\infty g}$ are $\mathcal{F}$-homotopy equivalent over B. We refer the reader to Theorem 3.2 and observe that the vertical homotopy $H\colon B \times I \to (B \sqcap_f E_\infty) * E_\infty$ of $s(f)$ into $s(g)$ is actually a vertical base homotopy ([6], Proposition 1.4). The argument used in (1.3) then shows that f and g are base homotopic. //

4 - Examples - In this section we develop examples where the previous general theory is applied.

Example 1 - Hurewicz fibrations - Let $\mathcal{F}_F$ be the category of spaces of the homotopy type of a fixed space F and whose morphisms are homotopy equivalences. We take $\mathcal{F}_F$ to be the admissible category consisting of Hurewicz fibrations over CW-complexes with fibres of the homotopy type of F; axiom A4 is a consequence of Corollary 7 [7].

Theorem 4.1 - Let $p_\infty\colon E_\infty \to B_\infty$ be a Hurewicz fibration with fibres of the homotopy type of F and B_∞ a CW-complex. Then p_∞ is Universal in all the four senses described in section 2 if, and only if, it is Universal in any one of these four

Proof - In view of Theorems 3.1, 3.2, 3.3 and the observation at the beginning of section 3, it is enough to show that if p_∞ is Free Universal then it is Aspherical Universal. Let f and c be respectively an arbitrary and the constant map of X into $\mathrm{Prin}_F E_\infty$. By (1.2) there are two pairs of fibre-preserving maps (f_1, f_0), (c_1, c_0) of the trivial fibration $pr_1 \colon X \times F \to X$ into p_∞, such that both f_0 and c_0 are classifying maps for pr_1. Lemma 1 of [10] implies that f_1 and c_1 are homotopic by a fibre-preserving homotopy over a homotopy $f_0 \simeq c_0$. Using (1.2) we conclude that $f \simeq c$. In other words, $[X, \mathrm{Prin}_F E_\infty] = 0$. The Proposition of [5] is now applied to the trivial fibration $X \times \mathrm{Prin}_F E_\infty \to X$ to show that $[X, \mathrm{Prin}_F E_\infty]_* = 0$ (the same result can be obtained from [12], 7.3.2). //

We observe at this point that Guy Allaud has shown in [1] that Grounded Universality for Hurewicz fibrations implies Aspherical Universality under the extra assumption that F has the homotopy type of a CW-complex; his proof involves the construction of auxiliary quasi-fibrations and the Brown representability theorem. In a subsequent paper [2], Allaud proved that if $\mathrm{Prin}_F E_\infty$ is contractible and F is locally compact then p_∞ is Grounded Universal. We will see in section 5 (5) that the conditions of 4.1 are equivalent to $\mathrm{Prin}_F E_\infty$ being contractible.

Example 2 - Principal G-bundles - Let G be a topological group. We use $\mathcal{F}_G$ to denote the category whose objects are right G-spaces Y such that, for all y in Y, the function

$$\tilde{y} \colon G \to Y, \quad \tilde{y}(g) = y \cdot g$$

is a homeomorphism, and whose morphisms are G-maps. We take F to be the right G-space. It is easily seen that the maps y are all G-homeomorphisms and that the G-maps from G to G are precisely the G-homeomorphisms $\tilde{a} \colon G \to G$; hence axiom (1.1) is satisfied.

We define $\mathcal{A}_G$ to be the category of principal G-bundles over CW-complexes. If $p \colon X \to A$ and $q \colon Y \to B$ are bundles in this category then the local triviality carries over to $p *_G q: X *_G Y \to A \times B$, using the weak topology on $X *_G Y$ with respect to this structure [8,7.5]. We see that $X *_G Y$ is a subspace of $X \cdot Y$ using the k-ified version of [4, example 2.3] and section 3a) of [7]; hence $\mathcal{A}_G$ is admissible.

Let $p: E \to B$ be a principal G-bundle. The function $\mathrm{Prin}_G E \to E$ which evaluates at the identity of G is a homemmorphism, its inverse being adjoint to the right action $E \times G \to E$; hence to say that p is Aspherical Universal means that E is aspherical.

Theorem 4.2 - Let $p_\infty: E_\infty \to B_\infty$ be a principal G-bundle, where B_∞ is a CW-complex. If p_∞ is Universal in any one of the four senses of this paper, or if E_∞ is contractible, then it satisfies each of these five conditions.

Proof - This follows from the results of section 3, [8, Theorem 7.5] and the observation preceeding the Theorem.

Example 3 - H-principal fibrations [9] - Let H be an H-space in the sense of being strictly associative, having a strict unit e and a homotopy inverse μ such that the composite

$$H \xrightarrow{\Delta} H \times H \xrightarrow{1 \times \mu} H \times H \xrightarrow{m} H$$

is homotopic to the constant map to e. A right action of H on a space X is a map $X \times H \to X$ such that $x(hh') = (xh)h'$ and $xe = x$.

Choose F to be the space H with the obvious right action on itself. Let $\mathcal{F}_H$ be the category whose objects are spaces X with a right action of H such that for all $y \in Y$,

$$\tilde{y}: H \to Y, \quad \tilde{y}(h) = y \cdot h$$

is a homotopy equivalence and whose morphisms are H-maps.

It is easily verified that the H-maps $g: H \to H$ are precisely the maps $\tilde{y}: H \to H$, and that these maps are H-homotopy equivalences. It is immediate that axiom (1.1) is satisfied.

We define $\mathcal{A}_H$ to be the category of H-principal fibrations in the sense of [9]. If $p: X \to A$ and $q: Y \to B$ are H-principal fibrations, then the H-fibre homotopy local triviality carries over to $p *_H q: X *_H Y \to A \times B$ (this follows using the k-ified version of [4, Theorem 1.1] and section 3a) of [7]. Hence $\mathcal{A}_H$ satisfies axiom A4 and so is admissible.

Theorem 4.3 - Let $p_\infty: E_\infty \to B_\infty$ be a H-principal fibration, where B_∞ is a CW-complex. If p_∞ is Universal in any one of the four senses of this paper, or if E_∞ is contractible, then it satisfies each of these five conditions.

This result is obtained by mimicking the proof of Theorem 4.2, but using the modified Dold-Lashof fibration p_H of [9] in our imitation of the appropriate part of [8, Theorem 7.5].

Example 4 - Trivial Fibrations - Let F and $\mathcal{F}_F$ be as in example 1. We consider the class $\mathcal{A}_T$ of all trivial fibrations over CW-complexes with fibre in $\mathcal{F}_F$, i.e., fibrations that are, to within a homeomorphism of their total spaces, projections of the product of their base space and a space in $\mathcal{F}_F$. It is clear that $\mathcal{A}_T$ is admissible and that the map $F \to *$ is both Grounded and Free Universal. It is clearly not (in general) Aspherical Universal (since $\mathrm{Prin}_F F$ is the space of self-homotopy equivalences of F). That it is not Extension Universal, follows either by Theorem 3.1 or more directly, because the evaluation map $e: H \times F \to F$ does not extend to $CH \times F$, where $H = H(F,F)$ is the space of self homotopy equivalences of F with the usual mapping space topology, and CH is the cone on H.

We have seen that in something weaker than an admissible category, Grounded Universality implies Free Universality; our next example shows that if the conditions on the category are further relaxed, then this theorem, may no longer hold.

Example 5 - Let $K(\pi,n+1)$ be a given Eilenberg-MacLane space $(n > 0)$. We take $F = K(\pi,n) = \Omega K(\pi,n+1)$ and $\mathcal{F}_S$ to be the category of homotopy equivalences between spaces having the homotopy type of $K(\pi,n)$. Let $\mathcal{A}_S$ denote the category of Hurewicz fibrations whose base spaces are simply-connected CW-complexes and whose fibres have the homotopy type of $K(\pi,n)$.

It follows from Postnikov factorization-type arguments that the path-fibration $K(\pi,n) \to PK(\pi,n+1) \to K(\pi,n+1)$ is Grounded Universal relative to this class. However the set of fibre homotopy equivalence classes of such fibrations is classified by $H^{n+1}(B,\pi)/\mathrm{Aut}\,\pi$ (see for example [3]), whilst $[B,K(\pi,n+1)] \cong H^{n+1}(B,\pi)$, so the path fibration is not Free Universal. Nor is it either Aspherical or Extension Universal, for if it were the arguments of this paper would ensure that it was Free Universal. The argument given at the beginning of section 3 that Grounded Universal implies Free Universal, works with a category of fibrations whose base spaces run over the class of all CW-complexes. The proof there depends on the fact that this class is closed under the operation of adjoining points. The class of simply-connected CW-complexes used here, does not satisfy this crucial condition.

5. Relations between Universal Hurewicz fibrations, Universal H-principal fibrations and Universal principal G-bundles - Theorems 4.1, 4.2 and 4.3 allow us to refer to these types of Universal fibrations in an unambiguous fashion.

(1) Given that p_∞ is Universal amongst Hurewicz fibrations with fibre of the homotopy type of F, then the associated $H(F,F)$ - principal fibration $\mathrm{prin}_F p_\infty$ has aspherical total space $\mathrm{Prin}_F E_\infty$, and by Theorem 4.3, $\mathrm{prin}_F p_\infty$ is Universal amongst $H(F,F)$ - principal fibrations (compare with [13, Theorem 3]).

(2) Given that G is a topological group and that $p_G': E_G' \to B_G'$ is a Universal principal G-bundle, the asphericity of E_G' and Theorem 4.3 ensures that p_G' is Universal amongst G-principal fibrations.

(3) Returning now to the general situation, we notice that a slight modification of the last part of Theorem 3.2 shows that any Aspherical Universal $\mathcal{A}$-fibration is actually a terminal object in the homotopy category of $\mathcal{A}$. Hence Aspherical Universal $\mathcal{A}$-fibrations are unique up to $\mathcal{F}$-homotopy equivalence.

(4) It follows that if p_∞ is Universal amongst Hurewicz fibrations with fibre F then $\mathrm{prin}_F p_\infty$ and the modified Dold-Lashof construction $p_{H(F,F)}$ of [9] are $\mathcal{F}_{H(F,F)}$-homotopy equivalent.

(5) Hence $\mathrm{Prin}_F E_\infty$ is a contractible space, generalizing Corollary 4.2 of [1].

(6) If G is a topological group, p_G' denotes the corresponding Milnor principal G-bundle and p_G the modified Dold-Lashof construction, then p_G and p_G' are $\mathcal{F}_G$-homotopy equivalent.

References

[1] Allaud, G. - On the Classification of Fiber Spaces. Math. Z. 92, 110-125 (1966).

[2] Allaud, G. - Concerning universal fibrations and a theorem of E. Fadell. Duke Math. J. 37, 213-224 (1970).

[3] Booth, P. - The Exponential law of maps, II. Math. Z. 121, 313-319 (1971).

[4] Booth, P. and Brown, R. - On the applications of fibered mapping spaces to exponential laws for bundles, ex-spaces and other categories of maps. To Appear.

[5] Booth, P., Heath, P. and Piccinini, R. - Section and Base-Point Functors. Math. Z. 144, 181-184 (1975).

[6] Booth, P., Heath, P. and Piccinini, R. - Restricted Homotopy Classes. An Acad. Brasil. Ci. 49 (1977), 1-8.

[7] Booth, P., Heath, P. and Piccinini, R. - Fibre preserving maps and functional spaces. These proceedings.

[8] Dold, A. - Partitions of Unity in the Theory of Fibrations. Annals of Math. 78, 223-255 (1963).

[9] Fuchs, M. - A modified Dold-Lashof Construction that does classify H-Principal Fibrations. Math. Ann. 192, 328-340 (1971).

[10] Gottlieb, D. - Correction to "On fibre spaces and the evaluation map". Annals of Math. 87, 640-642 (1968).

[11] May, J. P. - Classifying Spaces and Fibrations, Amer. Math. Soc. Memoirs no. 155 (1975).

[12] Spanier, E. - Algebraic Topology. New York: McGraw-Hill 1966.

[13] Stasheff, J. - _H-spaces and classifying spaces: foundations and applications._ Proc. Symp. Pure Math. XXII. Providence: Amer. Math. Soc. 1971.

[14] Steenrod, N. - The Topology of Fibre Bundles. Princeton, N.J.: Princeton Un. Press 1951.

ON ORBIT SETS FOR GROUP ACTIONS AND LOCALIZATION

Peter Hilton

To WF

. Introduction.

It is evident that, if one is to do homotopy theory for non-simply-connected paces, one should not restrict oneself to based spaces, based maps, and based omotopies, but should try to work in the *free* category. For the introduction of ase points, and the consequent based theory, justify themselves in the simply-onnected case since, on the one hand, the device is then purely technical, in the recise sense that, if X is 1-connected, then the free homotopy classes of maps $\to X$ coincide with the based homotopy classes of (based) maps $W \to X$, and, on he other hand, the introduction of base points enables group structure to be ntroduced naturally into the pointed set $[W,X]$ of based homotopy classes, at east for special classes of spaces X (grouplike spaces) or for special classes of spaces W (Cogrouplike spaces).

It is the intention of Guido Mislin, Joseph Roitberg and the author to make a systematic study of free maps and free homotopies, especially in the case in which X is nilpotent. Now, assuming W, X connected, the set of free homotopy classes of maps $W \to X$, which we write (W,X), is the orbit set of $[W,X]$ under the action of $\pi_1 X$. Thus, just as the study of $[W,X]$, with X nilpotent, depends on a preliminary study of nilpotent groups, so the study of (W,X) depends on a preliminary study of orbit sets for nilpotent actions of nilpotent groups Q on nilpotent groups N.

It is this study which is undertaken in the present series of lectures. We are concerned especially with questions of localization. We already know [1] that we may associate with a nilpotent action of the nilpotent group Q on the nilpotent group N an action (of no greater nilpotency class) of Q_P on N_P, where P is a family of primes and Q_P, N_P are the P-localizations of Q, N. One may ask whether the orbit set N_P/Q_P can lay any claim to being called the

P-localization of N/Q and, if so, what properties does the construction share with those of the localizations of nilpotent groups.

We show that the answer to the first question is positive by obtaining analogues, for orbit sets and their localizations, of certain basic theorems for nilpotent groups. All our results are motivated by our intended applications to the free homotopy theory of nilpotent spaces and these applications will appear in a subsequent paper by the three authors named above [5]. However, the results would surely seem to have a certain independent algebraic interest. Moreover, an unexpected formal connection appeared, as the work was proceeding, between the theory for orbit sets N/Q and that for *based* homotopy classes $[W,X]$. For if W is presented as the mapping cone of a map $S^{n-1} \to V$, $n \geq 2$, then $[W,X]$ has a natural structure as a disjoint union of abelian groups, each a homomorphic image of $\pi_n X$. Moreover, the restriction of e_*: $[W,X] \to [W,X_P]$ to each of these abelian groups is the P-localization homomorphism on that group. Our analogy is the following: if $\mathrm{nil}_Q N = c$ (see [1]) and if $\Gamma = \Gamma_Q^{c-1} N$ then Γ is central in N and N/Q has a natural structure as a disjoint union of abelian groups, each a homomorphic image of Γ. Moreover, the restriction of e_*: $N/Q \to N_P/Q_P$ to each of these abelian groups is the P-localization homomorphism on that group.

In Section 1 we develop this last point of view, using as our basic tool what we call the *exact orbit sequence* for the Q-action on N; actually there is one such sequence for each element $a \in N$. The sequence begins with maps which are homomorphisms and terminates with the fixed surjective function $\bar{\kappa}$: $N/Q \twoheadrightarrow M/Q$ induced by the projection κ: $N \twoheadrightarrow M = N/\Gamma$. It is by means of this sequence that we are able to pass from nilpotent group theory to the more general theory for orbit sets for nilpotent actions. The main theorem of this Section, apart from the existence of the orbit sequences, is a finiteness theorem for localization of orbit sets for nilpotent actions on *finitely generated* groups N (Theorem 1.2).

There is an important respect in which the localization theory for orbit sets differs from that for groups, and here again the analogue with the situation for the based homotopy sets $[W,X]$ is remarkably close--and not by coincidence, in view of the relationship established above. If N is a finitely generated nilpotent group, we can certainly find a *cofinite* family of primes P such that $N_P \to N_o$ is injective; we simply choose any P which excludes those primes p such that N has p-torsion. We show by an example that, even with Q and N finitely generated, we can even find an example such that $N_p/Q_p \to N_o/Q_o$ fails to be injective for every p; we recall that the example proposed by Adams and described in [4] had precisely this property, that is, W is compact, X is compact nilpotent, and $[W,X_p] \to [W,X_o]$ fails to be injective for every p. On the other hand--again the analogy with based homotopy theory is extremely close--for any nilpotent Q-action on a finitely generated nilpotent group N, and for any $x \in N_o/Q_o$, there exists a cofinite family P such that x lifts uniquely into N_P/Q_P (Theorem 1.7); the point to be emphasized is that P will depend on x.

In Section 3 we prove a Hasse Principle (the use of the name is due to Sullivan [6]) asserting essentially that one can recover all information about N/Q from its localizations N_p/Q_p at individual primes; and in Section 4 we prove a second basic pullback property for N/Q relating it to its *local expansion* [3] $\hat{N}/\hat{Q} = \prod_p N_p/Q_p$. Section 2 is brief and a little different in kind. It refers to *crossed homomorphisms* and is included because of its relevance to free homotopy theory and because it uses the technique of semidirect products which lies at the heart of Section 1.

The techniques developed here should certainly be applicable to a study of profinite completion (and p-profinite completion) in the case that Q, N are finitely generated. Such a study is also being undertaken by Guido Mislin, Joseph Roitberg and the author.

It is a pleasure to acknowledge frequent and very helpful conversations and exchange of letters with Guido Mislin and Joseph Roitberg. It is also a pleasure to achkowledge that the idea for the orbit sequence came from work of Richard Steiner who developed a very similar sequence in the special case of a group acting on itself by conjugation and was kind enough to show his idea to the author.

1. The exact orbit sequence.

In this section Q is a nilpotent group acting nilpotently on the group N. Thus N is nilpotent as a group and we recall from [1] that, if P is any family of primes, then there is an induced nilpotent action of Q_P and N_P, compatible with the given action of Q on N in the sense that the diagram

$$\begin{array}{ccc} Q & \longrightarrow & \text{Aut } N \\ \downarrow e & & \downarrow \\ Q_P & \longrightarrow & \text{Aut } N_P \end{array} \tag{1.1}$$

commutes. Indeed, we have the semi-direct product $N \land Q$ associated with the action of Q on N, and by localizing the right-split short exact sequence

$$N \rightarrowtail N \land Q \overset{\longleftarrow}{\twoheadrightarrow} Q \tag{1.2}$$

we obtain a right-split short exact sequence

$$N_P \rightarrowtail (N \land Q)_P \overset{\longleftarrow}{\twoheadrightarrow} Q_P, \tag{1.3}$$

identifying $(N \land Q)_P$ as $N_P \land Q_P$ for the given nilpotent action of Q_P on N_P.

Now let $a \in N$ and let $Q(a)$ be the subgroup of Q consisting of elements which fix a.

Theorem 1.1. $Q(a)_P = Q_P(ea)$, *where* $e\colon N \to N_P$ *P-localizes.*

Write N/Q for the set of orbits of N under the action of Q.

Theorem 1.2. *Let* N *be finitely generated and let* $S \subseteq T$ *be families of primes. Then* $N_T/Q_T \to N_S/Q_S$ *is finite-one.*

We will prove both these theorems by setting up a certain exact sequence. Assume that $\text{nil}_Q N = c$ and let $\Gamma = \Gamma_Q^{c-1} N$. Then

$$\Gamma \rightarrowtail N \overset{\kappa}{\twoheadrightarrow} M \quad (M = N/\Gamma) \tag{1.4}$$

is a central extension of Q-groups such that Q acts trivially on Γ and il$_Q M = c - 1$. We construct a sequence

$$Q(a) \rightarrowtail Q(\kappa a) \xrightarrow{\delta} \Gamma \xrightarrow{\rho} N/Q \xrightarrow{\bar{\kappa}} \twoheadrightarrow M/Q \tag{1.5}$$

s follows. First we embed $Q(a)$, in the obvious way, in $Q(\kappa a)$. We define δ y

$$\delta x = a(xa)^{-1}, \; x \in Q(\kappa a); \tag{1.6}$$

nd we define ρ by

$$\rho b = \overline{ba}, \; b \in \Gamma, \tag{1.7}$$

here $\bar{a}'$ is the orbit of $a' \in N$. Finally, $\bar{\kappa}$ is induced by κ.

roposition 1.3. *In the sequence* (1.5),

(i) δ *is a homomorphism with range* Γ, *and the sequence is exact at* $Q(\kappa a)$;

(ii) $\rho b = \rho b' \Leftrightarrow \exists x \in Q(\kappa a)$ *with* $b = b'(\delta x)$

(iii) $\rho\Gamma = \bar{\kappa}^{-1}(\overline{\kappa a})$.

roof. (i) It is plain from (1.6) that $\delta x = 1 \Leftrightarrow x \in Q(a)$. Moreover, since $\in Q(\kappa a)$, $\kappa\delta x = 1$ so that $\delta x \in \Gamma$. Finally δ is a homomorphism since

$$\begin{aligned} \delta(xy) = a(xya)^{-1} &= a(xa)^{-1}xa(xya)^{-1} \\ &= a(xa)^{-1}a(ya)^{-1}, \quad \text{since } Q \text{ operates trivially on } \Gamma, \\ &= (\delta x)(\delta y). \end{aligned}$$

(ii) It follows from (1.7) that $\rho b = \rho b' \Leftrightarrow \overline{ba} = \overline{b'a} \Leftrightarrow \exists x \in Q$ such that $'a = x(ba)$. But $x(ba) = b(xa)$ since Q operates trivially on Γ; and the quation $b'a = x(ba)$ forces $x \in Q(\kappa a)$, since $\kappa(ba) = \kappa(b'a) = \kappa a$. Thus $b = \rho b' \Leftrightarrow \exists x \in Q(\kappa a)$ such that $b(xa) = b'a$, or $b = b'(\delta x)$.

(iii) We have

$$\begin{aligned} \overline{\kappa a'} = \overline{\kappa a} &\Leftrightarrow \exists x \in Q, \text{ such that } \kappa a' = \kappa(xa) \\ &\Leftrightarrow \exists x \in Q, b \in \Gamma, \text{ such that } a' = b(xa) = x(ba) \\ &\Leftrightarrow \exists b \in \Gamma, \text{ such that } \overline{a'} = \overline{ba} \\ &\Leftrightarrow \overline{a'} \in \rho\Gamma. \end{aligned}$$

Corollary 1.4. *We may give the set* $\rho\Gamma$ *a unique (commutative) group structure so that* $\rho\colon \Gamma \twoheadrightarrow \rho\Gamma$ *is a homomorphism. In this group structure* $\bar{a}$ *is the neutral element.*

We call (1.5) *the exact orbit sequence of* (Q,N) *at* a. If we confine ourselves to

$$(1.8) \qquad \rho\Gamma \rightarrowtail N/Q \overset{\kappa}{\twoheadrightarrow} M/Q,$$

we have the *short orbit sequence at* a.

Note. In (1.5) and (1.8) it is necessary to emphasize that ρ *depends on* a. Thus, in (1.8), $\rho\Gamma$ is the *kernel* of $\bar{\kappa}$ in the sense that it is the $\bar{\kappa}$-counter-image of the base-orbit $\overline{\kappa a}$ of M/Q. We may write $\rho(a)$ for ρ.

We use the exact orbit sequence to prove Theorems 1.1 and 1.2.

Proof of Theorem 1.1. The diagram (1.1) induces the commutative diagram

$$(1.9) \qquad \begin{array}{ccccc} Q(a) & \rightarrowtail & Q(\kappa a) & \longrightarrow & \Gamma \\ \downarrow e_* & & \downarrow e_* & & \downarrow e \\ Q_P(ea) & \rightarrowtail & Q_P(\kappa_P ea) & \longrightarrow & \Gamma_P \end{array} ;$$

recall that (1.4) behaves well under localization, in the sense that localization is exact and, indeed for all i,

$$(\Gamma_Q^i N)_P = \Gamma_{Q_P}^i (N_P). \quad \text{(Theorem 3.2 of [1])}$$

We now argue by induction on c, the Q-nilpotency class of N. If $c = 1$, then $Q(a) = Q$, $Q_P(ea) = Q_P$, so that $e_*\colon Q(a) \to Q_P(ea)$ certainly localizes. Assume inductively that the theorem holds for $c - 1$. Then $e_*\colon Q(\kappa a) \to Q_P(\kappa_P ea) = Q_P(e\kappa a)$ localizes. It then follows from (1.9) and the exactness of localization that $e_*\colon Q(a) \to Q_P(ea)$ localizes.

Corollary 1.5. *Let localization induce the commutative diagram*

$$(1.10)\qquad \begin{array}{ccc} N/Q & \xrightarrow{\bar\kappa} & M/Q \\ \downarrow e_* & & \downarrow e_* \\ N_P/Q_P & \xrightarrow{\bar\kappa_P} & M_P/Q_P \end{array}$$

Then, for each $a \in N$, *the restriction of* $e_*: N/Q \to N_P/Q_P$ *to* $\rho(a)\Gamma$ *induces the localizing map*

$$e: \rho(a)\Gamma \to \rho_P(ea)\Gamma_P = (\rho(a)\Gamma)_P.$$

Proof. We extend (1.9) to the right by

$$\begin{array}{ccccccc} Q(\kappa a) & \longrightarrow & \Gamma & \twoheadrightarrow & \rho\Gamma & \rightarrowtail & N/Q \\ \downarrow e & & \downarrow e & & \downarrow & & \downarrow e_* \\ Q_P(\kappa_P ea) & \longrightarrow & \Gamma_P & - & \rho_P\Gamma_P & \rightarrowtail & N_P/Q_P \end{array}$$

and invoke Theorem 1.1 and the exactness of localization.

Proof of Theorem 1.2. If N is finitely generated, so are Γ and M and we argue by induction on $c = \text{nil}_Q N$. If $c = 1$, then $N_P/Q_P = N_P$, for all families P, and $N_T \to N_S$ is certainly finite-one, being a homomorphism whose kernel is the (T-S)-torsion of N. Our inductive hypothesis will imply that $M_T/Q_T \to M_S/Q_S$ is finite-one. Consider (see (1.10))

$$(1.11)\qquad \begin{array}{ccc} N_T/Q_T & \xrightarrow{\bar\kappa_T} & M_T/Q_T \\ \downarrow e_* & & \downarrow e_* \\ N_S/Q_S & \xrightarrow{\bar\kappa_S} & M_S/Q_S \end{array}$$

Choose $u \in e_*(N_T/Q_T)$. Then $\bar\kappa_S u$ has finitely many e_*-counterimages. Thus to prove the theorem, it suffices to show that if $u \in N_S/Q_S$, $v \in M_T/Q_T$ with $\bar\kappa_S u = e_* v$, then there are only finitely many $w \in N_T/Q_T$ with $e_* w = u$, $\bar\kappa_T w = v$. If there is no such w, we have no problem. If there is such a w, choose a particular $w = \bar a$, $a \in N_T$, and extend (1.11) to the left to construct a map of short orbit sequences

$$\begin{array}{cccccl} \rho_T\Gamma_T & \rightarrowtail & N_T/Q_T & \xrightarrow{\bar\kappa_T} & M_T/Q_T, & \quad \rho_T = \rho_T(a) \\ \downarrow e & & \downarrow e_* & & \downarrow e_* & \\ \rho_S\Gamma_S & \rightarrowtail & N_S/Q_S & \xrightarrow{\bar\kappa_S} & M_S/Q_S, & \quad \rho_S = \rho_S(ea) \end{array}$$

Then any other candidate for w, say $\bar{a}'$, lies in $\rho_T\Gamma_T$ and, by Corollary 1.5, the set of all possible w is the kernel of the localizing map $e\colon \rho_T\Gamma_T \to \rho_S\Gamma_S$. Since Γ is finitely generated this kernel is finite,[1] so the theorem is proved.

Remark. We may regard Γ as acting on N/Q by $b.\bar{a}' = \overline{ba'}$, $b \in \Gamma$, $\bar{a}' \in N/Q$. Then, by (1.7) and the exactness of (1.5), the isotropy subgroup of $\bar{a}$ under this action is $\delta Q(\kappa a)$. Thus $\rho\Gamma$ is faithfully represented on $\bar{a}$, in the sense that there is a bijection (indeed, an identity) between the orbit of $\bar{a}$ under Γ and $\rho\Gamma$. It follows that N/Q has the structure of a disjoint union of Γ-orbits,

$$(1.2) \qquad N/Q = \coprod_{\Gamma\text{-orbits}} \rho(a)\Gamma$$

Moreover, each $\rho(a)\Gamma$ may be given a canonical commutative group structure and then the function $e_*\colon N/Q \to N_P/Q_P$, induced by localization, maps Γ-orbits to Γ_P-orbits and indeed localizes each Γ-orbit to the appropriate Γ_P-orbit. Thus N/Q has a natural structure richer than that of a set and $e_*\colon N/Q \to N_P/Q_P$ may be regarded as a localization map with respect to that richer structure. The situation here is very much analogous to that of the based homotopy set $[W,X]$ (see [4]), when the CW-complex W is regarded as the mapping cone of an attaching map $S^{n-1} \to V$, $n \geq 2$.

In [4] we elaborated the counterexample suggested by Adams to show that, even with W finite and X nilpotent of finite type, one cannot guarantee the existence of a *cofinite* family of primes P such that $[W,X_P] \to [W,X_0]$ is injective. Indeed, in that example, $[W,X_p] \to [W,X_0]$ fails to be injective for *every* prime p -- and this despite the fact, referred to above, that $[W,X]$ is a disjoint union of commutative groups each of which P-localizes under the map induced by P-localization $X \to X_P$. We may give a similar (but simpler) counterexample to show that there exist Q, N finitely generated such that $N_p/Q_p \to N_0/Q_0$ fails to be injective for every p; thus, although N/Q has a 'group-like'

[1] The argument is exactly as for the corresponding step of Theorem 1.7, so we do not give it twice. Here we have written $\rho_T\Gamma_T$, for aesthetic reasons, although there is no suggestion that a is in the image of $e\colon N \to N_T$ and that therefore $\rho_T\Gamma_T$ arises from T-localization of some $\rho\Gamma$.

tructure it still fails to behave precisely like a group.

xample 1.6. Let Q be cyclic infinite, generated by x; and let N be free belian (as a group) on two generators (u,v). We write N *additively* and rescribe the Q-action by

$$xu = u, \quad xv = u + v. \tag{1.13}$$

hen $\mathrm{nil}_Q N = 2$, Γ is cyclic infinite, generated by u, and M is cyclic nfinite, generated by $v \bmod \Gamma$. Q acts trivially on Γ and M.

Let $a = kv$, for some integer k. If $k \neq 0$, then $Q(a) = 1$, $Q(\kappa a) = Q$, nd $\delta: Q \to \Gamma$ is given by $\delta x = a - xa = -ku$. Thus $\rho(a)\Gamma$ is cyclic of order $|k|$. If $k = 0$ then, of course $\rho(a)\Gamma = \rho(0)\Gamma$ is cyclic infinite. Moreover, y varying k, we run through all the Γ-orbits of N/Q, each appearing once only, o that

$$N/Q = \coprod_{k \in \mathbb{Z}} \mathbb{Z}/|k|. \qquad (\mathbb{Z}/0 = \mathbb{Z})$$

t is now plain from the previous discussion that N_p/Q_p contains many (disjoint) opies of $\mathbb{Z}/p$, one for each $k \neq 0$ such that $p|k$, and that each copy is nnihilated under $N_p/Q_p \to N_o/Q_o$, which cannot therefore be injective.

In view of the observation, in our Remark, on the analogy between the structure of the based homotopy set $[W,X]$ and that of the orbit set N/Q, we would expect not only to find an analogue of the Adams counterexample but also an analogue of the positive assertion Theorem 2.10 of [4]. This we now present.

heorem 1.7. *If the nilpotent group* Q *acts nilpotently on the finitely generated group* N *and if* $x \in N_o/Q_o$, *then there exists a cofinite family of primes* P *such that* x *lifts uniquely into* N_S/Q_S *for all* $S \subseteq P$.

Proof. We argue by induction on $c = \mathrm{nil}_Q N$, the result being well known if $c = 1$ (see, e.g., Theorem 2.9 of [4]). Also it is plain, from nilpotent group theory, that there certainly exists a cofinite family P_1 such that x lifts to N_S/Q_S for $S \subseteq P_1$. We invoke the exact sequence (1.4),

$$\Gamma \rightarrowtail N \overset{\kappa}{\twoheadrightarrow} M,$$

and assume inductively that there exists a cofinite family P_2 such that $\bar{\kappa}_o x$

lifts uniquely into M_S/Q_S for all $S \subseteq P_2$. Set $R = P_1 \cap P_2$ and let $\bar{a} \in N_R/Q_R$ map to $x \in N_o/Q_o$.

Now it follows from Theorem 1.1 that $Q_R(a)$ is R-local. It thus further follows from the exactness of (1.5) that[1] $\rho(a)\Gamma_R$ is R-local; as a homomorphic image of Γ_R it is a finitely generated $\mathbb{Z}_R$-module. We now form the map of short orbit sequences,

$$\begin{array}{ccccc} \rho(a)\Gamma_R & \rightarrowtail & N_R/Q_R & \twoheadrightarrow & M_R/Q_R \\ \downarrow e & & \downarrow & & \downarrow \\ \rho_o(ea)\Gamma_o & \rightarrowtail & N_o/Q_o & \twoheadrightarrow & M_o/Q_o \end{array} \tag{1.14}$$

The first arrow in (1.14) is rationalization (Corollary 1.5). Thus the kernel of $e\colon \rho(a)\Gamma_R \to \rho_o(ea)\Gamma_o$ is finite, being the torsion part of the finitely generated $\mathbb{Z}_R$-module $\rho(a)\Gamma_R$. Let T be the *finite* family of primes p such that $\rho(a)\Gamma_R$ has p-torsion, let T' be the complement of T, and let $P = R \cap T'$. Then P is cofinite and if $S \subseteq P$ then (1.14) factors as

$$\begin{array}{ccccc} \rho(a)\Gamma_R & \rightarrowtail & N_R/Q_R & \overset{\bar{\kappa}_R}{\twoheadrightarrow} & M_R/Q_R \\ \downarrow & & | & & \downarrow \\ \rho_S(e_1a)\Gamma_S & \rightarrowtail & N_S/Q_S & \overset{\bar{\kappa}_S}{\twoheadrightarrow} & M_S/Q_S \\ \downarrow & & | & & \downarrow \\ \rho_o(ea)\Gamma_o & \rightarrowtail & N_o/Q_o & \overset{\bar{\kappa}_o}{\twoheadrightarrow} & M_o/Q_o \end{array} \tag{1.15}$$

where $e_1\colon N_R \to N_S$, $e_2\colon N_2 \to N_o$, $e = e_2e_1$. Moreover, by our choice of S, $\rho_S(e_1a)\Gamma \to \rho_o(ea)\Gamma_o$ is injective. We also know, by our inductive hypothesis, that $\bar{\kappa}_o x$ lifts uniquely into M_S/Q_S; and that x lifts to $\overline{e_1a}$ in N_S/Q_S. It now follows that $\overline{e_1a}$, together with any other lift of x, lies in $\rho_S(e_1a)\Gamma_S$ and all lifts have the same image in $\rho_o(ea)\Gamma_o$. Thus the lift of x to N_S/Q_S is unique, and the inductive step is complete.

---

[1] Note that we write $\rho(a)$, not $\rho_R(a)$, since there is no reason to suppose that $\rho(a)$ arises by R-localization.

. On crossed homomorphisms.

Again, Q is a nilpotent group acting nilpotently on the group N. A unction $\delta:Q \rightarrow N$ satisfying the condition

$$\delta(xy) = (\delta x)(x\delta y),\ x,\ y \in Q, \tag{2.1}$$

s called a *crossed homomorphism*. The following proposition is surely well-nown[1].

roposition 2.1. *Let* Q *act on* N *and form the semi-direct product* $N \rtimes Q$ *and he right-split, short exact sequence*

$$N \overset{\iota}{\rightarrowtail} N \rtimes Q \underset{\sigma}{\overset{\pi}{\rightleftarrows}} Q; \tag{2.2}$$

se ι, σ *to embed* N, Q *in* $N \rtimes Q$. *Then there is a natural equivalence* $\longleftrightarrow d$ *between crossed homomorphisms* $\delta: Q \to N$ *and homomorphisms* $d: Q \to N \rtimes Q$ *uch that* $\pi d = 1$, *given by*

$$dx = (\delta x, x). \tag{2.3}$$

oreover, under this natural equivalence, we have

$$\ker \delta = d^{-1}Q. \tag{2.4}$$

roof. If δ is a crossed homomorphism then $d(xy) = (\delta(xy), xy) = ((\delta x)(x\delta y), xy) = \delta x, x)(\delta y, y) = (dx)(dy)$, so that d is a homomorphism. The converse is equally bvious; just as obvious is the naturality of the equivalence. Finally, if $x = 1$, then $dx = (1,x) \in Q$; and if $dx = (1,y)$, $y \in Q$, then $\delta x = 1$, $x = y$ o that $x \in \ker \delta$.

This proposition enables us to prove a result on the localization of crossed omomorphisms which will be useful in the topological applications.

heorem 2.2. *Let the nilpotent group* Q *act nilpotently on the group* N, *let be a family of primes, and let us consider the associated* Q_P*-action on* N_P *in he sense of Section 1. Then, with every crossed homomorphism* $\delta: Q \to N$ *we may*

---

[1] Of course, no nilpotency assumption is required in this proposition.

associate a unique crossed homomorphism $\delta_p: Q_p \to N_p$ *such that the diagram*

$$(2.5)\qquad \begin{array}{ccc} Q & \xrightarrow{\delta} & N \\ {\scriptstyle e}\downarrow & & \downarrow{\scriptstyle e} \\ Q_p & \xrightarrow{\delta_p} & N_p \end{array}$$

commutes. Moreover $(\ker \delta)_p = \ker \delta_p$.

Proof. Construct $d: Q \to N \wedge Q$ with $\pi d = 1$, as in Proposition 2.1. According to (1.2), (1.3), we may then localize d to $d_p: Q_p \to N_p \wedge Q_p$ and $\pi_p d_p = 1$. Again by Proposition 2.1 we associate with d_p a crossed homomorphism $\delta_p: Q_p \to N_p$. Since $d_p e = (e \wedge e)d$, it follows that $\delta_p e = e\delta$, so that (2.5) commutes. Conversely, if δ_p were any crossed homomorphism making (2.5) commutative, then the associated homomorphism $d_p: Q_p \to N_p \wedge Q_p$ would satisfy $d_p e = (e \wedge e)d$ and would thus be uniquely determined; δ_p is therefore itself uniquely determined by the commutativity of (2.5).

Since (Theorem I.2.10 of [2]) localization commutes with pull-backs, it follows that, if $K = \ker \delta = d^{-1}Q$, then $K_p = d_p^{-1}Q_p = \ker \delta_p$, and the theorem is proved.

Remark. There is an evident generalization of Proposition 2.1 and Theorem 2.2, where R is a nilpotent group, $\alpha: R \to Q$ is a homomorphism and $\delta: R \to N$ is said to be α-*crossed* if $\delta(xy) = (\delta x)(\alpha x.\delta y)$, $x, y \in R$. The associated $d: R \to N \wedge Q$ is then given by $dx = (\delta x, \alpha x)$.

3. The Hasse Principle for orbit sets.

Our objective is again to prepare the way for a proof of the Hasse Principle for free homotopy classes. In fact, we establish the necessary intermediate step between group theory and free homotopy theory.

Again, Q is a nilpotent group operating nilpotently on the group N. We have maps $e_p: N/Q \to N_p/Q_p$, $r_p: N_p/Q_p \to N_o/Q_o$ induced by localization and rationalization, with $r_p e_p$ independent of p.

heorem 3.1. *Let* $\bar{a}, \bar{a}' \in N/Q$ *be such that* $e_p\bar{a} = e_p\bar{a}'$ *for all* p. *Then* $\bar{a} = \bar{a}'$.

roof. We set $c = \mathrm{nil}_Q N$ and argue by induction on c. If $c = 1$ the result is well-known (see p. 28 of [2]). For the inductive step we construct the sequence (1.4),

$$\Gamma \rightarrowtail N \twoheadrightarrow M \tag{3.1}$$

and assume the corresponding conclusion for M. Thus we have

$$\begin{array}{ccc} N/Q & \xrightarrow{\bar{\kappa}} & M/Q \\ \downarrow \hat{e} & & \downarrow \hat{e}, \\ \prod_p N_p/Q_p & \xrightarrow{\prod_p \bar{\kappa}_p} & \prod_p M_p/Q_p \end{array} \qquad \hat{e} = \{e_p\}, \tag{3.2}$$

and we wish to prove the left-hand $\hat{e}$ injective. Since $\hat{e}\bar{a} = \hat{e}\bar{a}'$ it follows from our inductive hypothesis that $\bar{\kappa}\bar{a} = \bar{\kappa}\bar{a}'$. Thus we may complete (3.2) to a map of short orbit sequences

$$\begin{array}{ccccc} \rho\Gamma & \rightarrowtail & N/Q & \twoheadrightarrow & M/Q \\ \downarrow \hat{e} & & \downarrow \hat{e} & & \downarrow \hat{e} \\ \prod_p \rho_p\Gamma_p & \rightarrowtail & \prod_p N_p/Q_p & \twoheadrightarrow & \prod_p M_p/Q_p \end{array} \quad \begin{array}{l} , \ \rho = \rho(a), \\ \\ , \ \rho_p = \rho_p(e_pa) \end{array} \tag{3.3}$$

and $\bar{a}, \bar{a}' \in \rho\Gamma$ with $\hat{e}\bar{a} = \hat{e}\bar{a}'$. But, on the left of (3.3), we have the standard group-theoretical version of $\hat{e}$, known to be injective, so that $\bar{a} = \bar{a}'$ as required.

Corollary 3.2. *Let* N *be a nilpotent group. Then two elements of* N *are conjugate if and only if their images in every localization* N_p *are conjugate.*

Proof. We allow N to act on itself by conjugation. Since N is nilpotent, this action is nilpotent; and the elements of the orbit set of N under this action are the conjugacy classes.

We now turn to the second half of the Hasse Principle.

Theorem 3.3. *Let* N *be finitely generated and let* $x_p \in N_p/Q_p$ *be elements such that* $r_p x_p = r_q x_q$ *for all primes* p, q. *Then there exists a unique* $x \in N/Q$ *such that* $e_p x = x_p$ *for all* p.

Proof. We have only to prove the existence of x and we again argue by induction on $c = \text{nil}_Q N$. If $c = 1$ this is Theorem I.3.6 of [2]. For the inductive step we construct (3.1) and assume the corresponding conclusion for M. Then it follows from (3.2), since $r_p \bar{\kappa}_p = \bar{\kappa}_o r_p$, that there exists a unique $y \in M/Q$ with $e_p y = \bar{\kappa}_p x_p$ for all p. Let $y = \kappa \bar{a}$, $a \in N$, and complete (3.2) to (3.3) as above, with $\rho = \rho(a)$. Since, for all p, $\bar{\kappa}_p x_p = \bar{\kappa}_p e_p \bar{a}$, it follows that $x_p \in \rho_p \Gamma_p$, for all p. Moreover $r_p x_p$, as an element of $\rho_o \Gamma_o$, is independent of p. Thus we may invoke the standard group-theoretical Hasse Principle to infer the existence of x (in fact, in $\rho\Gamma$), such that $e_p x = x_p$ for all p.

Remarks. 1. By developing an appropriate localization theory for such generalized commutative group structures as we discussed in the Remark in Section 1 (see (1.12)), it should be possible to derive the Hasse Principle of this Section very rapidly.

2. It is commented in [2] that we may generalize the Hasse Principle (for nilpotent groups) by replacing the partitioning of the set of all primes into the individual primes by an arbitrary partitioning of Π. Moreover the restriction that N be finitely generated in the existence part of the Hasse Principle is not necessary if Π is partitioned as a *finite* union of disjoint subsets. Precisely the same generalization is available with regard to Theorems 3.1 and 3.3.

4. A pullback diagram.

Theorem I.3.7 of [2] asserts that, if N is finitely generated nilpotent and if $\hat{N} = \prod_p N_p$, then the diagram

$$\begin{array}{ccc} N & \xrightarrow{\hat{e}} & \hat{N} \\ \downarrow r & & \downarrow r \\ N_o & \xrightarrow{e_o} & \hat{N}_o \end{array} \qquad (4.1)$$

where $\hat{e} = \{e_p\}$ and r is rationalization, is a pullback. Theorem 3.1 of [3] asserts that (4.1) remains a pullback even if N is not finitely generated. We propose to generalize this to orbit sets.[1]

We first need to prepare the ground with a general remark. We know, as in Section 1, that we may associate with a nilpotent action of the nilpotent group Q on the group N, a nilpotent action of Q_P on N_P. This association passes to morphisms in the following obvious sense.

<u>Proposition 4.1</u>. *Let* $\phi\colon N \to N'$ *be a map of* Q*-groups. Then, for any family of primes* P, $\phi_P\colon N_P \to N'_P$ *is a map of* Q_P*-groups.*

In fact, we will use a generalization of this. If $\alpha\colon Q \to Q'$, $\phi\colon N \to N'$ are homomorphisms and if N is a Q-group, N' a Q'-group, then ϕ is an α-*map* if

$$\phi(xa) = (\alpha x)(\phi a). \tag{4.2}$$

Then Proposition 4.1 generalizes to

<u>Proposition 4.2</u>. *Let* $\alpha\colon Q \to Q'$, *and let* $\phi\colon N \to N'$ *be an* α*-map, where* N *is* Q*-nilpotent,* N' *is* Q'*-nilpotent. Then* $\phi_P\colon N_P \to N'_P$ *is an* α_P*-map.*

<u>Proof</u>. The homomorphism ϕ is an α-map if and only if $\phi \wedge \alpha : N \wedge Q \to N' \wedge Q'$ is a homomorphism.

Now $e \wedge e\colon N \wedge Q \to N_P \wedge Q_P$ P-localizes. Thus we have a homomorphism $\tau\colon N_P \wedge Q_P \to N'_P \wedge Q'_P$, uniquely determined by the commutativity of

$$\begin{array}{ccc} N\wedge Q & \xrightarrow{\phi\wedge\alpha} & N'\wedge Q' \\ \downarrow{\scriptstyle e\wedge e} & & \downarrow{\scriptstyle e\wedge e} \\ N_P\wedge Q_P & \xrightarrow{\tau} & N'_P\wedge Q'_P \end{array} \tag{4.3}$$

and we have only to show that

$$\tau = \phi_P \wedge \alpha_P, \tag{4.4}$$

[1] Actually Theorem 3.1 of [3] asserts that (4.1) is bicartesian, but we cannot expect to generalize the pushout property of (4.1) to orbit sets.

to infer that ϕ_p is an α_p-map. Now let $\iota : N \rightarrowtail N \rtimes Q$, $\sigma : Q \rightarrowtail N \rtimes Q$ be the canonical embeddings in the semidirect product. Then

$$\tau(e\wedge e)\iota = \tau\iota e, \quad (e\wedge e)(\phi\wedge\alpha) = \iota e\phi = \iota\phi_p e,$$

so that $\tau\iota e = \iota\phi_p e: N \to N'_p \wedge Q'_p$. Since $N'_p \wedge Q'_p$ is P-local, the universality of e implies

$$\tau\iota = \iota\phi_p,$$

or

$$\tau(a,1) = (\phi_p a,1), \; a \in N_p. \tag{4.5}$$

Similarly, by composing $\tau(e\wedge e)$ and $(e\wedge e)(\phi\wedge\alpha)$ with σ, we infer that

$$\tau(1,x) = (1,\alpha_p x), \; x \in Q_p. \tag{4.6}$$

Thus $\tau(a,x) = \tau(a,1)\tau(1,x) = (\phi_p a,1)(1,\alpha_p x) = (\phi_p a,\alpha_p x) = (\phi_p\wedge\alpha_p)(a,x)$, and (4.4) is proved.

We now return to the generalization of (4.1). Given the Q-nilpotent group N (with Q itself nilpotent), then the *local expansion*, in the language of [3], $\hat{e}: N \to \hat{N}$ is an $\hat{e}$-map with respect to the local expansion $\hat{e}: Q \to \hat{Q}$. There is thus an induced function of orbit sets which we will also write $\hat{e}: N/Q \to \hat{N}/\hat{Q}$; indeed, $\hat{N}/\hat{Q} = \prod_p N_p/Q_p$, and $\hat{e}$ is the function discussed in Section 3. However, our preparatory remarks above show that we may *rationalize* $\hat{e}: N/Q \to \hat{N}/\hat{Q}$. For the local expansion $\hat{e}: N \to \hat{N}$ being an $\hat{e}$-map with respect to $\hat{e}: Q \to \hat{Q}$, it follows that $\hat{e}_o: N_o \to \hat{N}_o$ is an $\hat{e}_o$-map with respect to $\hat{e}_o: Q_o \to \hat{Q}_o$. Thus we may also form the map of orbit sets,

$$\hat{e}_o: N_o/Q_o \to \hat{N}_o/\hat{Q}_o.$$

Plainly the diagram

$$\begin{array}{ccc} N/Q & \xrightarrow{\hat{e}} & \hat{N}/\hat{Q} \\ \downarrow r & & \downarrow r \\ N_o/Q_o & \xrightarrow{\hat{e}_o} & \hat{N}_o/\hat{Q}_o \end{array} \tag{4.3}$$

s commutative. By Theorem 3.1 we know that $\hat{e}$ is injective.

heorem 4.3. *The diagram (4.3) is a pullback.*

roof. We have only to prove *existence*, in the sense that if $\bar{x} \in \hat{N}/\hat{Q}$, $\in N_o/Q_o$ with $r\bar{x} = \hat{e}_o\bar{y}$, then there exists $\bar{z} \in N/Q$ with $\hat{e}\bar{z} = \bar{x}$, $r\bar{z} = \bar{y}$. or, by Theorem 3.1, $\hat{e}$ is injective, so the *uniqueness* of $\bar{z}$ follows immediately.

Now if x, y are in the orbits of $\bar{x}, \bar{y}$, then there exists $u \in \hat{Q}_o$ with $rx = u.\hat{e}_o y$. By Proposition 3.4 of [3], $u = (rv)(\hat{e}_o w)$, $v \in \hat{Q}$, $w \in Q_o$. Thus

$$r(v^{-1}x) = \hat{e}_o(wy),$$

so that, by Theorem 3.1 of [3], there exists $z \in N$ with $\hat{e}z = v^{-1}x$, $rz = wy$; but then $\hat{e}\bar{z} = \bar{x}$, $r\bar{z} = \bar{y}$, and the theorem is proved.

References

1. Peter Hilton, Nilpotent actions on nilpotent groups, Proc. Logic and Math. Conference, Springer Lecture Notes, 450 (1975), 174-196.
2. Peter Hilton, Guido Mislin and Joseph Roitberg, *Localization of Nilpotent Groups and Spaces*. Mathematics Studies 15, North Holland (1975).
3. Peter Hilton and Guido Mislin, Bicartesian squares of nilpotent groups, Comm. Math. Helv. 50 (1975), 477-491.
4. Peter Hilton, Guido Mislin and Joseph Roitberg, On maps of finite complexes into nilpotent spaces of finite type: a correction to 'Homotopical Localization', Proc. London Math. Soc. (1977).
5. Peter Hilton, Guido Mislin and Joseph Roitberg, On free maps and free homotopies (to appear).
6. Dennis Sullivan, Genetics of homotopy theory and the Adams conjecture, Ann. of Math. 100 (1974), 1-79.

ON FREE MAPS AND FREE HOMOTOPIES INTO NILPOTENT SPACES

Peter Hilton, Guido Mislin, Joseph Roitberg and Richard Steiner[1]

1. Introduction

So long as the techniques of homotopy theory were largely being applied to simply-connected spaces, it was eminently reasonable to place oneself in the homotopy category of based spaces and based homotopy classes of based maps. For, by this device, one would very often have group structure in the set of based homotopy classes $[W,X]$--in fact, if X is grouplike or W is cogrouplike--so that there would be algebraic tools readily available; and there would be no loss of generality since, if X is 1-connected, there is a bijection between $[W,X]$ and (W,X), the set of free homotopy classes of free maps.

However, considerable attention has been given in recent years to the homotopy theory of nilpotent spaces and it would therefore seem that the time is appropriate to study *free* maps and *free* homotopies of maps of spaces W into nilpotent spaces X. If X is connected, there is a surjection $\theta\colon [W,X] \to (W,X)$; indeed, (W,X) may be identified with the orbit-set of $[W,X]$ under the action of $\pi_1 X$. Thus an important distinction arises between the based and the free theories. Moreover, if W is a suspension then (W,X) has the structure of an orbit-set for a group action on a group. This algebraic situation was discussed in [H] with a view to its applications to the study of (W,X).

It should be pointed out that the concept of nilpotency is itself free. That is, although it is defined in terms of homotopy groups it is independent of the choice of base point. Thus it is perfectly meaningful to say that a connected space is nilpotent. Similarly, the universal property of the P-localization, where P is a family of primes, while originally stated in terms of based homotopy, also turns out to be a free concept. By this we mean the following. Let $e\colon X \to X_P$ be P-localization, and let $f\colon X \to Y$ be a map to a P-local space. Then there exists a map $g\colon X_P \to Y$ such that ge is freely homotopic to f, and g is determined up to free homotopy. To see this, we first remark that we may

[1]The last named author was supported by the National Research Council of Canada.

endow our spaces with 'secret' base points (a technique used frequently in this paper) and we may assume that $f(o) = \dot{o}$. Then we know that there exists $g: X_p \to Y$ such that ge is based-homotopic to f. Now if $h: X_p \to Y$ is such that he is freely homotopic to f, we may suppose $h(o) = \dot{o}$; then there exists $\xi \in \pi_1 Y$ such that $\xi(he)$ is based-homotopic to f. But $\xi(he) = (\xi h)e$, so that g is based-homotopic to ξh and hence freely homotopic to h.

These considerations suggest that we should regard the localization theory of nilpotent spaces as a free theory and attempt to establish the basic theorems of the theory to parallel the development of the based theory. We begin this program in the present paper[1]. In fact, our principal theorems are the following, where we write WX for the free function space, to distinguish it from the based function space X^W. Note that every free function space WX may be regarded as a based function space X^{W^+}, where W^+ is obtained from W by taking the disjoint union with a base point; of course W^+ is not connected. However, our theorems will not require the connectedness of the domain complex. Thus the passage to the free theory may also be regarded as generalization, to non-connected W, of the based theory.

Theorem A *Let* X *be nilpotent and* W *homologically finite. Then each component of* WX *is nilpotent.*

Now let $f: W \to X$ be a map; write (WX, f), (X^W, f) for the component of the function space containing f.

Theorem B *Let* X *be nilpotent and* W *homologically finite. Then* $(WX, f)_P = (WX_P, ef)$.

Theorem C *Let* W *be a homologically finite complex, let* X *be nilpotent of finite type and let* $S \subseteq T$ *be families of primes. Then (i)* $e_*: (W, X_T) \to (W, X_S)$ *is finite-one; and (ii)* $e_*: [W, X_T] \to [W, X_S]$ *is finite-one.*

Our next theorem enunciates the local Hasse Principle for free homotopy. We

---

[1] Some remarks about the free theory were made in [HMR], but they were essentially perfunctory.

state it in two parts, a uniqueness part and an existence part, since the existence part requires stronger hypotheses. Note that the homotopy relation is *free* homotopy, though the statements are, of course, also true for based homotopy.

Theorem D1 *Let* W *be a homologically finite complex, let* X *be nilpotent, and let* $f, g\colon W \to X$ *be two maps such that* $e_p f \simeq e_p g\colon W \to X_p$ *for all primes* p. *Then* $f \simeq g$.

Theorem D2 *Let* W *be a homologically finite complex, let* X *be nilpotent of finite type, and let* $f(p)\colon W \to X_p$ *be maps, one for each prime* p, *such that* $r_p f(p)\colon W \to X_0$ *is, up to homotopy, independent of* p. *Then there exists a map* $f\colon W \to X$ *such that* $e_p f \simeq f(p)$ *for all* p. (Here $r_p\colon X_p \to X_0$ is rationalization.)

Our last principal theorem is concerned with what was called the *local expansion* in [HM]. Thus, given a nilpotent space X, we construct $\check{X}$ which is a CW-model for ΠX_p and there is an evident map $\check{e}\colon X \to \check{X}$. It is known (see [HM]) that the square

$$\begin{array}{ccc} [W,X] & \xrightarrow{\check{e}} & [W,\check{X}] \\ \downarrow r & & \downarrow r \\ [W,X_0] & \xrightarrow{\check{e}_0} & [W,\check{X}_0] \end{array} \qquad (1.1)$$

is cartesian if W is finite connected. We prove

Theorem E *If* W *is homologically finite and* X *is nilpotent, then (1.1) and*

$$\begin{array}{ccc} (W,X) & \xrightarrow{\check{e}} & (W,\check{X}) \\ \downarrow r & & \downarrow r \\ (W,X_0) & \xrightarrow{\check{e}_0} & (W,\check{X}_0) \end{array} \qquad (1.2)$$

are cartesian.

All these theorems follow immediately from results in [H] if W is a suspension.

With regard to the methods used in this paper, a remark of a general nature about free homotopy theory is in order. A standard technique of based homotopy heory is that of exploiting the familiar duality which yields essentially equiva-ent approaches to a given problem of the theory. This duality fails in the free heory. The basic reason for this failure is that, whereas the product of two paces remains the same whether we work in the free or based category, the co-roduct changes, being the disjoint union in the free theory and the 'wedge' in he based theory. Thus we continue to have group structure in (W,X) if X is rouplike--indeed, then, X being connected, $(W,X) = [W,X]$--but we do not get roup structure in (W,X) if W is cogrouplike. Further, in a fibration (typical f a refined Postnikov tower of a nilpotent space X)

$$\begin{array}{ccc} K(G,n) & \longrightarrow & Y & & \\ & & \downarrow & & \\ & & Z & \longrightarrow & K(G,n+1) \end{array}$$

e have an operation of $H^n(W;G)$ on (W,Y); but, in a cofibration (typical of a ellular decomposition of a connected space W)

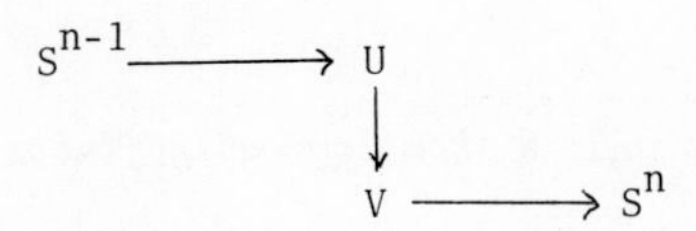

e do not have an operation of $\pi_n X$ on (V,X).

The effect of this failure of duality is that, in the free theory, we have wo methods which are related but distinct. In this paper we have consistently dopted the former approach (via the refined Postnikov tower of X). The superiority of this approach for the problem under consideration in this paper had already been ffectively demonstrated by one of us (see remarks in [St2] and the appendix to St1]). In particular, Steiner has pointed out (a) with this approach it is not ecessary to treat the case of W connected seperately, and (b) it is sufficient o assume W *homologically* finite (that is, the integral homology of W is globally finitely generated), whereas the 'dual' method seems to require, in its most easily accessible form, that W be actually of the homotopy type of a finite complex.[1] It is hoped to devote a later paper to a more systematic comparison to the two approaches in free homotopy theory.

The approach via the refined Postnikov tower of X can be used, as in [S], to study *completion*, rather than localization, of nilpotent spaces of finite type, since completion respects fibrations of such spaces. The advantages are the same as for localization. By a merely formal change of argument we are able to prove the appropriate variants of THEOREMS B, D, E, which we collect together in an appendix.

It seems plain that one would wish to follow up this beginning by applying and generalizing the methods to a study of the homotopy theory of (nilpotent) group actions on spaces which move the base point--based on a study of affine action of groups. These investigations are currently being undertaken by the authors.

2. The nilpotency of the free function space

We prove

THEOREM A *Let* X *be nilpotent and* W *homologically finite. Then every component of the free function space* WX *is nilpotent.*

Proof. We may suppose that X has a finite Postnikov tower, since W is homologically finite, and we argue by induction on the height of a refined principal tower for X. Now if X is an Eilenberg-MacLane space $K(G,m)$ with G abelian then WX is an (abelian) grouplike space and hence certainly all its components are nilpotent. This starts the induction; to handle the inductive step we consider the fibration

$$(2.1)\qquad \begin{array}{ccc} K(G,n) & \longrightarrow & Y \\ & & \downarrow q \\ & & Z \xrightarrow{\ h\ } K(G,n+1) \end{array}$$

and pick $f\colon W \to Y$ to specify a component (WY,f) of WY. We may suppose $(WZ,\bar{f})$

[1] Thus the approach adopted in this paper has advantages for the based theory as well.

nilpotent where $\bar{f} = qf$; but

$$WY \longrightarrow WZ \xrightarrow{Wh} WK \tag{2.2}$$

is a fibration $(K = K(G,n+1))$. Let $\pi \subseteq \pi_1(WK,o)$ be the image of $\pi_1(WZ,\bar{f})$ under Wh, and let L_π be the covering space of (WK,o) with fundamental group π. It is plain that (2.2) induces a fibration

$$(WY,f) \longrightarrow (WZ,\bar{f}) \longrightarrow L_\pi, \tag{2.3}$$

so that the nilpotency of (WY,f) follows from Theorem II.2.2 of [HMR]. This completes the inductive step.

For THEOREM B in the next section we will need a refinement of the observation that WK is an (abelian) grouplike space if $K = K(G,m)$ with G abelian. Consider the fibration

$$K^W \rightarrowtail WK \xrightarrow{ev} K.$$

This is a fibration of abelian grouplike spaces with an evident cross-section, given by the subgroup of WK consisting of the constant maps. Moreover, this cross-section is, of course, a homomorphism. We infer

Theorem 2.1 *If* $K = K(G,m)$ *with* G *abelian, then there is an isomorphism of abelian grouplike spaces*

$$WK \cong K^W \times K.$$

Note THEOREM A was stated in [HMR] but not used. The based version with W connected was proved--and used.

3. Localization of free function spaces

We place ourselves in the situation of THEOREM A, so that X is a nilpotent space, W is a homologically finite complex and $f: W \to X$ is a map which specifies a component (WX,f) of the free function space WX and provides a base-point for this component. We know from THEOREM A that (WX,f) is nilpotent. Now let P be a family of prime numbers and let $e: X \to X_P$ be P-localization. We will prove THEOREM B by a proof-strategy similar to that used for THEOREM A.

THEOREM B *Let* X *be nilpotent and* W *homologically finite. Then* $(WX,f)_P = (WX_P,ef)$.

Proof. We may suppose that X has a finite Postnikov tower, since W is homologically finite, and we argue by induction on the height of a refined principal tower for X. Moreover it is plain that we may assume in the proof that W is connected. Now if X is an Eilenberg-MacLane space $K(G,m)$ with G abelian then it follows from Theorem 2.1 that

$$(WX,f)_P = (X^W,f)_P \times X_P = (X_P^W,ef) \times X_P = (WX_P,ef).$$

To handle the inductive step, we again consider the fibration

$$\begin{array}{ccc} K(G,n) & \longrightarrow & Y \\ & & \downarrow q \\ & & Z \longrightarrow K(G,n+1) \end{array} \tag{3.1}$$

and assume that $(WZ,\bar{f})_P = (WZ_P,e\bar{f})$ where $f: W \to Y$ and $\bar{f} = qf$. Then (3.1) P-localizes to

$$\begin{array}{ccc} K(G_P,n) & \longrightarrow & Y_P \\ & & \downarrow q_P \\ & & Z_P \longrightarrow K(G_P,n+1) \end{array} \tag{3.1$_P$}$$

We thus have an induced map of the fibration $WY \to WZ \to WK(G,n+1)$ into the fibration $WY_P \to WZ_P \to WK(G_P,n+1)$ and hence, for each $i \geq 1$, a map of exact sequences (with $K = K(G,n+1)$)

$$\begin{array}{ccccccccc} \pi_{i+1}(WZ,\bar{f}) & \longrightarrow & \pi_{i+1}(WK,o) & \longrightarrow & \pi_i(WY,f) & \longrightarrow & \pi_i(WZ,\bar{f}) & \longrightarrow & \pi_i(WK,o) \\ \downarrow e_1 & & \downarrow e_2 & & \downarrow e_3 & & \downarrow e_4 & & \downarrow e_5 \\ \pi_{i+1}(WZ_P,e\bar{f}) & \longrightarrow & \pi_{i+1}(WK_P,o) & \longrightarrow & \pi_i(WY_P,ef) & \longrightarrow & \pi_i(WZ_P,e\bar{f}) & \longrightarrow & \pi_i(WK_P,o) \end{array}, \tag{3.2}$$

where e_1, e_2, e_4, e_5 P-localize. It follows that e_3 P-localizes and the inductive step is complete.

We may use THEOREM B to study the *stabilizer* of f in $\pi_1 X$. Since $\pi_1 X$ operates on $[W,X]$ we may define the subgroup stab f of $\pi_1 X$ to be the isotropy subgroup of $f \in [W,X]$. Let X_f be the covering space of X corresponding to this subgroup; it is then easy to see that, if $ev\colon WX \to X$ is the evaluation map, then $ev|(WX,f)$ lifts to ev' such that

$$(W^X, f) \longrightarrow (WX, f) \xrightarrow{ev'} X_f \tag{3.3}$$

is a fibration of connected spaces. Moreover, if X is nilpotent and W homologically finite then (3.3) is a fibration of nilpotent spaces. Thus we obtain the corollary of THEOREM B:

<u>Corollary 3.1</u> *Under the hypotheses of* THEOREM B, $(\text{stab } f)_P = \text{stab } ef$, *where* $e\colon X \to X_P$.

<u>Note</u>. (i) A special case of THEOREM B was proved in [R]. THEOREM B was stated in [HMR] but not used. The based version, with W connected, was proved--and used.

(ii) In Theorem 6.1 of [HR] we included an addendum relating to free function spaces. The proof of the addendum, along the lines of the argument given there, would require the following assertion about stab f.

<u>Proposition 3.2</u> *If* $\mathcal{C}$ *is a proper Serre class of nilpotent groups and* $g\colon X \to X'$ *is a* $\mathcal{C}$-*equivalence, then* $g_*\colon \text{stab } f \to \text{stab } gf$ *is a* $\mathcal{C}$-*bijection.*

This proposition will be proved as a consequence of Theorem 4.3 below. However, we could avoid invoking this proposition by proving both the based and the free versions of Theorem 6.1 of [HR] by adopting the dual approach of this paper, since one may always assume that a map of nilpotent spaces $g\colon X \to X'$ sends a refined principal Postnikov tower of X to a refined principal Postnikov tower of X'. Indeed, if we adopt the canonical refinement introduced in the proof of Theorem II.2.9 of [HMR; p. 66] then this follows immediately from the characteristic property of the lower central series of a group or of a π-module.

4. A finiteness theorem

In this section we propose to prove the analogue of Corollary II.5.4(a) of [HMR]. Thus we must study the function

$$e_*: (W, X_T) \to (W, X_S)$$

induced by localization $e: X_T \to X_S$, where $S \subseteq T$ are families of primes; and our theorem (THEOREM C below) will assert that e_* is finite-one if W is a homologically finite CW-complex and X is of finite type.

We argue as in the previous sections. Thus, in the refined Postnikov tower of X, we meet an induced fibration

(4.1)
$$\begin{array}{ccc} K(G,n) \subseteq Y & & \\ \downarrow q & & \\ Z & \longrightarrow & K(G,n+1) \end{array}$$

where, of course,

(4.2) $q_*: \pi_1 Y \to \pi_1 Z$ *is an isomorphism if* $n \geq 2$ *and a surjection if* $n=1$.

Let $f: W \to Y$, $\bar{f} = qf: W \to Z$, where we suppose f to be base-point-preserving; we now regard (f), $(\bar{f})$ as elements of (W,Y), (W,Z), respectively.

The fibration (4.1) gives rise to a fibration of free function spaces

$$WK(G,n) \to WY \to WZ \to WK(G,n+1)$$

and hence to an exact sequence (compare (3.2))

(4.3) $$\ldots \to \pi_1(WZ, \bar{f}) \xrightarrow{\phi} \pi_1(WK(G,n+1), o) \to (W,Y) \xrightarrow{q_*} (W,Z).$$

<u>Proposition 4.1</u> $\pi_1(WK(G,n+1)) = H^n(W;G)$.

<u>Proof</u>. Write $K = K(G,n+1)$. Then, by Theorem 2.1, $WK \cong K^W \times K$ and K is 1-connected. This shows that $\pi_1(K^W) \to \pi_1(WK)$ is an isomorphism; and of course,

$_1(K^W) = H^n(W;G)$.

The standard facts of the homotopy theory of fibrations (compare the proof f Theorem II.5.3 of [HMR] now ensure

heorem 4.2 *In (4.3),* coker ϕ *operates faithfully on* $(f) \in (W,Y)$ *and there is n induced bijection*

$$\text{coker } \phi \cong q_*^{-1}(\bar{f}).$$

We may now prove the main result of this section.

HEOREM C *Let* W *be a homologically finite complex, let* X *be nilpotent of inite type and let* $S \subseteq T$ *be families of primes. Then (i)* $e_*: (W,X_T) \to (W,X_S)$ *s finite-one; and (ii)* $e_*: [W,X_T] \to [W,X_S]$ *is finite-one.*

roof. We will be content to prove (i)--the proof of (ii) is entirely analogous.[1] Je construct the refined Postnikov tower of X; since W is homologically finite, e may, without real loss of generality, suppose this tower to be finite. Thus we ay argue by induction on the height of the tower, starting with a point, for hich the assertion is trivial.

To carry out the inductive step, we revert to (4.1) and assume $(W,Z_T) \to (W,Z_S)$ o be finite-one. If we look at

$$\begin{array}{ccc} (W,Y_T) & \xrightarrow{q_{T*}} & (W,Z_T) \\ \downarrow e_* & & \downarrow e_* \\ (W,Y_S) & \xrightarrow{q_{S*}} & (W,Z_S) \end{array}$$

it is plain, from our inductive assumption, that all we have to prove is that, given $(f) \in (W,Y_T)$, there are only finitely many $(g) \in (W,Y_T)$ with $q_{T*}(g) = q_{T*}(f)$, $e_*(g) = e_*(f)$. Now, by Theorem 4.2, the set of elements (g) with $q_{T*}(g) = q_{T*}(f)$ is in natural bijective correspondence with coker ϕ_T, where

[1] As pointed out in the Introduction, (i) may be regarded as a special case of (ii); but we prefer to emphasize free homotopy.

$$\phi_T: \pi_1(WZ_T,\bar{f}) \to H^n(W;G_T).$$

Moreover, since we also insist that $e_*(g) = e_*(f)$, it is plain that the set of suitable elements (g) is in natural bijective correspondence with the kernel of the homomorphism

$$(4.4) \qquad e_*: \text{coker } \phi_T \to \text{coker } \phi_S,$$

induced by the localization $e: X_T \to X_S$.

Now since W is finite, $H^n(W;G_T) \to H^n(W;G_S)$ is localization. By THEOREM B, $\pi_1(WZ_T,\bar{f}) \to \pi_1(WZ_S,e\bar{f})$ is localization. Thus e_* in (4.4) is itself localization. Further, it follows from THEOREM B that $(WZ_T,\bar{f})$ is T-local; so therefore is $\pi_1(WZ_T,\bar{f})$ and hence coker ϕ_T. Moreover, since W is finite and X is of finite type, G is finitely generated and $H^n(W;G_T)$ is a finitely generated $\mathbb{Z}_T$-module; so therefore is coker ϕ_T. Thus the kernel of e_* in (4.4) is the (T-S)-torsion of a finitely generated $\mathbb{Z}_T$-module and hence finite.

This completes the inductive step and hence established the theorem.

<u>Remark</u> Theorem 1.2 in [H] immediately established THEOREM C when W is a suspension, with no recourse to an inductive argument or Theorem 4.1. For then $\pi_1 S$ is a nilpotent group operating nilpotently on the group $[W,X]$, which is finitely generated.

We close this section by establishing the relationship between Theorem 4.2 and the based version; in this way we will elucidate the nature of the homomorphism $\phi: \pi_1(WZ,\bar{f}) \to H^n(W;G)$. The key diagram, based on $f: W \to Y$, is

$$(4.5) \qquad \begin{array}{ccccccc} \pi_1(Z^W,\bar{f}) & \xrightarrow{\psi} & H^n(W;G) & \longrightarrow & [W,Y] & \longrightarrow & [W,Z] \\ \downarrow & & \| & & \downarrow & & \downarrow \\ \pi_1(WZ,\bar{f}) & \xrightarrow{\phi} & H^n(W;G) & \longrightarrow & (W,Y) & \longrightarrow & (W,Z) \end{array}$$

We have $q_*: \pi_1 Y \twoheadrightarrow \pi_1 Z$ and, by restriction

$$q_*: q_*^{-1} \text{ stab } \bar{f} \twoheadrightarrow \text{ stab } \bar{f}.$$

Moreover, stab $f \subseteq q_*^{-1}$ stab $\bar{f}$. Now there is a function $\delta: q_*^{-1}$ stab $\bar{f} \to$ coker ψ, given by $\delta\xi = \alpha$, where

$$\xi f = \alpha f, \ \xi \in q_*^{-1} \text{ stab } \bar{f}, \ \alpha \in \text{coker } \psi.$$

Since the operations of $\pi_1 Y$ and $H^n(W;G)$ on $[W,Y]$ commute, it follows readily that δ is a homomorphism; and it is plain that stab f is the kernel of δ.

Theorem 4.3 *There is an exact sequence*

$$\text{stab } f \rightarrowtail q_*^{-1} \text{ stab } f \xrightarrow{\delta} \text{coker } \psi \xrightarrow{\theta} \text{coker } \phi, \tag{4.6}$$

where θ *is induced by (4.5).*

Proof. We have only to prove exactness at coker ψ. Now, in (4.5),

$$\theta u = \theta v \Leftrightarrow u = \xi v, \ u, v \in [W,Y], \ \xi \in \pi_1 Y, \text{ and}$$
$$\theta(\alpha f) = (\theta\alpha)(f), \ \alpha \in \text{coker } \psi.$$

Thus if $\alpha = \delta\xi$, then $(\theta\alpha)(f) = \theta(\xi f) = \theta(f) = (f)$, so that $\theta\alpha$ is the neutral element of coker ϕ. Conversely, if $(\theta\alpha)(f) = (f)$, then $\theta(\alpha f) = (f)$, $\alpha f = \xi f$, $\xi \in \pi_1 Y$, and $(q_*\xi)\bar{f} = \bar{f}$, so that $\xi \in q_*^{-1}$ stab $\bar{f}$. From this it follows that $\alpha = \delta\xi$.

Plainly the sequence (4.6) localizes in the expected manner. Moreover, it may be used to provide a proof of Proposition 3.2, as promised. For, given a map $g: X \to X'$ of nilpotent spaces, we may always assume that g maps a refined Postnikov tower of X to a refined Postnikov tower of X'. Indeed, to justify this assumption, we may--and, in fact, will--take the canonical refinements constructed in the proof of Theorem II.2.9 of [HMR]. Thus we may assume that g is a map

$$\begin{array}{ccccccc} K(G,n) & \longrightarrow & Y & \xrightarrow{q} & Z & \longrightarrow & K(G,n+1) \\ \downarrow g & & \downarrow g & & \downarrow g & & \downarrow g \\ K(G',n) & \longrightarrow & Y' & \xrightarrow{q'} & Z' & \longrightarrow & K(G',n+1) \end{array} \tag{4.7}$$

where $G = \Gamma^i\pi_n X/\Gamma^{i+1}\pi_n X$, $G' = \Gamma^i\pi_n X'/\Gamma^{i+1}\pi_n X'$, for some i. Thus if $g: X \to X'$

is a $\mathcal{C}$-equivalence, for some proper Serre class $\mathcal{C}$ [HR], so is $g\colon K(G,m) \to K(G',m)$. From (4.7) and Theorem 4.3 we obtain a map of exact sequences

$$\begin{array}{ccccccc} \text{stab } f & \rightarrowtail & q_*^{-1}\ \text{stab } \bar{f} & \longrightarrow & \text{coker } \psi & \twoheadrightarrow & \text{coker } \phi \\ \downarrow g_1 & & \downarrow g_2 & & \downarrow g_3 & & \downarrow g_4 \\ \text{stab } gf & \rightarrowtail & q_*'^{-1}\ \text{stab } g\bar{f} & \longrightarrow & \text{coker } \psi' & \twoheadrightarrow & \text{coker } \phi' \end{array} \tag{4.8}$$

It follows from Theorem 6.1 of [HR] (the based version suffices) that g_3 is $\mathcal{C}$-bijective. Let us assume inductively that g induces a $\mathcal{C}$-bijective $g_*\colon \text{stab } \bar{f} \to \text{stab } g\bar{f}$. If $n > 1$ in (4.7), it follows at once that g_2 is $\mathcal{C}$-bijective. If $n = 1$, we invoke the map of short exact sequences

$$\begin{array}{ccccc} G & \rightarrowtail & q_*^{-1}\ \text{stab } \bar{f} & \twoheadrightarrow & \text{stab } \bar{f} \\ \downarrow g_* & & \downarrow g_2 & & \downarrow g_* \\ G' & \rightarrowtail & q_*'^{-1}\ \text{stab } g\bar{f} & \twoheadrightarrow & \text{stab } g\bar{f} \end{array}$$

to infer that g_2 is $\mathcal{C}$-bijective. Thus, in any case, it follows from (4.8) that g_1 is $\mathcal{C}$-bijective, and the inductive step in the proof of Proposition 3.2 is complete.

As pointed out in Section 3, Proposition 3.2 yields the free version of Theorem 6.1 [HR]; but we may, in fact, obtain both the based and the free version by adopting the 'dual' point of view of this paper.

5. The local Hasse principle in free homotopy

We prove here the analogue in free homotopy of Theorem II.5.1 of [HMR]; see also [S] and [St 1]--the free theory is treated in the appendix to the latter. We first prove the uniqueness part of the Hasse principle; note that the homotopy relation below is *free* homotopy.

Theorem D1 *Let* W *be a homologically finite complex, let* X *be nilpotent, and let* $f, g\colon W \to X$ *be two maps such that* $e_p f \simeq e_p g\colon W \to X_p$, *for all primes* p. *Then* $f \simeq g$.

Proof. We again argue by induction on the height of the refined Postnikov tower of X, which we may assume finite since W is homologically finite. Thus the essential step is the following: we have

$$\begin{array}{ccc} (W,Y) & \xrightarrow{q_*} & (W,Z) \\ \downarrow \check{e} & & \downarrow \check{e} \\ \prod_p (W,Y_p) & \xrightarrow{\Pi q_{p*}} & \prod_p (W,Z_p) \end{array}$$

where, by our inductive hypothesis, the right hand vertical arrow is injective, and we wish to prove the left hand vertical arrow injective. If $\check{e}(f) = \check{e}(g)$, where (f), (g) are the classes of f, g in (W,Y), then $q_*(f) = q_*(g)$, so that (g) belongs, essentially, to the group $H^f = \text{coker}\,\phi$, $\phi\colon \pi_1(WZ,\bar{f}) \to H^n(W;G)$, by Theorem 4.2. Moreover, as shown in the proof of THEOREM C, H^f localizes to $H^{e_p f}$ at the prime p, so that $\check{e}|H^f$ is the local expansion [HM], $\check{e}\colon H^f \to \prod_p H^{e_p f}$. But the local expansion is injective, so that (f) = (g) since $\check{e}(f) = \check{e}(g)$. This completes the proof.

We now turn to the existence part of the Hasse principle.

Theorem D2 *Let* W *be a homologically finite complex, let* X *be nilpotent of finite type, and let* $f(p)\colon W \to X_p$ *be maps, one for each prime* p, *such that* $r_p f(p)\colon W \to X_o$ *is, up to homotopy, independent of* p. *Then there exists a map* $f\colon W \to X$ *such that* $e_p f \simeq f(p)$ *for all* p.

Proof. Once again, we argue by induction on the height of the refined Postnikov tower of X. Thus we have, in an evident notation,

$$\begin{array}{ccc} (W,Y) & \xrightarrow{q_*} & (W,Z) \\ \downarrow \check{e} & & \downarrow \check{e} \\ \prod_p (W,Y_p) & \xrightarrow{\Pi q_p{}^*} & \prod_p (W,Z_p) \\ \downarrow r & & \downarrow r \\ (W,Y_o) & \xrightarrow{q_{o*}} & (W,Z_o) \end{array}$$

(here r stands for the *collection* of maps r_p)

By our inductive hypothesis, there exists a (unique) (h) ∈ (W,Z) such that

$$\check{e}(h) = \{q_{p*}(f(p))\} .$$

It is easy to see (by applying the Hasse principle to $H^{n+1}(W;G)$ and its localizations $H^{n+1}(W;G_p)$ -- note that X is of finite type, so that G is finitely generated) that (h) lifts to some $(g) \in (W,Y)$. Then, for each p, $q_{p*}(f(p)) = q_{p*}(e_p g)$. In the notation used in the proof of Theorem D1, we thus have

$$\begin{array}{ccc}
H^g & \subseteq & (W,Y) \\
\downarrow \check{e} & & \downarrow \check{e} \\
\prod_p H^{e_p g} & \subseteq & \prod_p (W,Y_p) \\
\downarrow r & & \downarrow r \\
H^{e_o g} & \subseteq & (W,Y_o)
\end{array}$$

and $f(p) \in H^{e_p g}$ with $r_p f(p) \in H^{e_o g}$ independent of p. Thus, by the Hasse principle for finitely generated abelian groups (note that H^g is finitely generated since X is of finite type), there exists $(f) \in H^g$ with $\check{e}(f) = \{f(p)\}$, and the inductive step is complete.

<u>Remark</u> As for THEOREM C, we may immediately infer THEOREM D, for W a suspension, from Theorem 3.1, 3.3 of [H].

6. A certain cartesian square

Let X be a nilpotent space and let X be its local expansion. Thus X is a CW-model for $\prod_p X_p$. We then have a commutative square

$$(6.1) \qquad \begin{array}{ccc}
X & \xrightarrow{\check{e}} & \check{X} \\
\downarrow r & & \downarrow r \\
X_o & \xrightarrow{\check{e}_o} & \check{X}_o
\end{array}$$

and we know (see [HM]) that, if W is finite connected, then

$$(6.2) \qquad \begin{array}{ccc}
[W,X] & \xrightarrow{\check{e}} & [W,\check{X}] \\
\downarrow r & & \downarrow r \\
[W,X_o] & \xrightarrow{\check{e}_o} & [W,\check{X}_o]
\end{array}$$

s a pullback. We now assert the free version. In fact, we claim

heorem E *If* W *is homologically finite and* X *is nilpotent, then (6.2) and*

$$\begin{array}{ccc} (W,X) & \xrightarrow{\check{e}} & (W,\check{X}) \\ \downarrow r & & \downarrow r \\ (W,X_0) & \xrightarrow{\check{e}_0} & (W,\check{X}_0) \end{array} \tag{6.3}$$

re cartesian.

roof. We do not need to enter into great detail. We assume X to have finite ostnikov tower and we argue by our usual induction to prove (6.3). Since we know hat $\check{e}: (W,X) \to (W,\check{X})$ is injective, we have only to show that if $a \in (W,\check{X})$, $\in (W,X_0)$, with $ra = \check{e}_0 b$, then there exists $(f) \in (W,X)$ with $\check{e}(f) = a$, $(f) = b$. The steps of the argument follow very closely those of the proof of HEOREM D2.

Further, we may remark, as in the previous two sections, that no new proof of HEOREM E is required if W is a suspension--we merely quote Theorem 4.3 of [H].

7. Free homotopy and completion

In this appendix we point out that we may handle the profinite-completion functor (or the p-profinite completion functor) just as we handled the localization functor in the previous sections, provided we assume that X is a nilpotent space of finite type. For we know that completion is compatible with fibrations, so we may complete the refined Postnikov tower of X. We do not need to go into details but may announce the theorems as follows.

Let $X^\wedge$ be the profinite completion of X and $X^\wedge_{(p)}$ the p-profinite completion of X. Of course, $X^\wedge = \prod_p X^\wedge_{(p)}$.

Theorem 7.1 *Let* X *be nilpotent of finite type and* W *homologically finite. Then if* $e^\wedge: X \to X^\wedge$, $e^\wedge_{(p)}: X \to X^\wedge_{(p)}$ *are completion maps, and if* $f: W \to X$,

$$(WX,f)^\wedge = (WX^\wedge, e^\wedge f), \quad (WX,f)^\wedge_{(p)} = (WX^\wedge_{(p)}, e^\wedge_{(p)} f).$$

Theorem 7.2 (Hasse principle for profinite completion) *Let* X *be nilpotent finite type and* W *homologically finite.*

If $f, g\colon W \to X$ *are such that* $e^{\wedge}_{(p)} f \quad e^{\wedge}_{(p)} g\colon W \to X^{\wedge}_{(p)}$ *for all* p, *ther* $f \simeq g$;

Note that there is no point in stating an existence part to the Hasse principle for profinite completion, since this forms part of the next theorem.

Theorem 7.3 *Let* X *be nilpotent of finite type and let* W *be homologically finite. Then the square*

$$\begin{array}{ccc} (W,X) & \xrightarrow{\;e^{\wedge}\;} & (W,X^{\wedge}) \\ \downarrow{\scriptstyle r} & & \downarrow{\scriptstyle r} \\ (W,X_0) & \xrightarrow{\;(e^{\wedge})_0\;} & (W,(X^{\wedge})_0) \end{array}$$

is cartesian.

References

[H] Peter Hilton, On orbit sets for group actions and localization (these Proceedings).

[HM] Peter Hilton and Guido Mislin, Remarkable squares of homotopy types, B Soc. Mat. Bras. 5 (1974), 165-180.

[HMR] Peter Hilton, Guido Mislin and Joseph Roitberg, *Localization of nilpot groups and spaces*, Math. Studies 15, North Holland (1975).

[HR] Peter Hilton and Joseph Roitberg, Generalized C-theory and torsion phe in nilpotent spaces, Houston Journal of Mathematics 2 (1976), 525-559.

[R] Jospeh Roitberg, Note on nilpotent spaces and localization, Math. Zeit (1974), 67-74.

[St 1] Richard Steiner, Localization, completion and infinite complexes, Mathematika 24 (1977), 1-15.

[St 2] Richard Steiner, Exact sequences of conjugacy classes and rationalizat Math. Proc. Cam. Phil. Soc. 82 (1977), 249-253.

[S] Dennis Sullivan, Genetics of homotopy theory and the Adams conjecture, Ann. of Math. 100 (1974), 1-79.

CONDITIONS FOR FINITE DOMINATION FOR CERTAIN COMPLEXES

Guido Mislin

In order to apply Wall's obstruction theory effectively, it is important to be able to prove that a space is dominated by a finite complex, or, more generally, that a space is of finite type. We will investigate these problems in some particular cases of non-nilpotent spaces, generalizing known results on nilpotent spaces (for a nilpotent space X the following holds: X is of finite type if and only if $H_i X$ is finitely generated for all i, and X is finitely dominated if and only if $\oplus H_i X$ is finitely generated, see [5]). The methods we use rely on results of K. Brown, P. Kahn and E. Dror ([2], [3], [4]).

1. Spaces of finite type

Let X be a connected CW-complex. We will say that X is of finite type, if X is homotopy equivalent to a complex with finite skeleta. A basic result of K. Brown states that X is of finite type if and only if $\pi_1 X$ is finitely presented and the functors $H_i(X,-)$, regarded as functors on the category of $\pi_1 X$-modules, commute with (arbitrary) products (cf. [2]).

Lemma 1.1. Let $F \to E \xrightarrow{p} B$ be a fibration of connected spaces with F and B of finite type. Then E is of finite type.

Proof. From the Serre spectral sequence with local coefficients

$$H_i(B,H_j(F,-)) \Rightarrow H_{i+j}(E,-)$$

one infers that $H_k(E,-)$ commutes with product, if the functors $H_j(F,-)$ and $H_i(B,-)$ do. It remains to show that $\pi_1 E$ is finitely presented. Without loss of generality we may assume that B is a finite complex, say $B = C \cup e^n$. Then $E = p^{-1}(C) \cup (e^n \times F)$ and it becomes obvious by induction on the number of cells of B and by applying van Kampen's theorem, that $\pi_1 E$ is finitely presented. Hence E is of finite type by Brown's result.

Corollary 1.2. Let $\bar{X} \to X$ be a (not necessarily regular) finite covering. Then X is of finite type if and only if $\bar{X}$ is.

Proof. Clearly, if X is of finite type then so is $\bar{X}$. Suppose now that $\bar{X}$ is of finite type. Choose a finite covering $Y \to \bar{X}$ such that $Y \to X$ is regular (e.g. Y the covering associated with $\bigcap g\pi_1\bar{X}g^{-1}$, $g \in \pi_1 X$). Then Y is of finite type, being a finite covering of $\bar{X}$. Consider now the fibration $Y \to X \to K(G,1)$ with $G = \pi_1 X/\pi_1 Y$. Then Y and $K(G,1)$ are both of finite type and so is X, by the previous Lemma.

Recall that X is called homologically nilpotent if $\pi_1 X$ operates nilpotently on $H_i\tilde{X}$ for all i, $\tilde{X}$ the universal covering of X. Furthermore, a group π is said to be of type $\overline{FP}$, if $\mathbb{Z}$ regarded as a trivial π-module, possesses a projective resolution consisting in finitely generated modules over $\mathbb{Z}\pi$. For instance, groups with noetherian group ring are of course of type $\overline{FP}$.

Theorem 1.3. Let X be a homologically nilpotent space with finitely presented fundamental group of type $\overline{FP}$. Then the following are equivalent.

(i) X is of finite type.

(ii) $\pi_i X$ is finitely generated for all $i \geq 2$.

(iii) $H_i X$ is finitely generated for all $i \geq 2$.

Proof. The equivalence of (ii) and (iii) under the conditions stated was established in [3, Prop. 5]. To see that (ii) implies (i) we consider the fibration $\tilde{X} \to X \to K(\pi_1 X,1)$. Clearly, (ii) implies that X is of finite type and, since $\pi_1 X$ is finitely presented and of type $\overline{FP}$, $K(\pi_1 X,1)$ is of finite type [2]. Hence X is of finite type by 1.1.

In view of Corollary 1.2 it is then clear that, for X to be of finite type, it suffices that X has a finite covering $\bar{X}$ for which the hypotheses of Theorem 1.3 together with one of the conditions (i), (ii) or (iii) hold.

For the considerations in the next section we will need the following lemma.

emma 1.4. Let X be a space of finite type and let $\{M_i, f_i : M_{i+1} \to M_i | i \in N\}$ enote an inverse system of $\pi_1 X$-modules with $\varprojlim^1 M_i = 0$. Then there is a short xact sequence

$$0 \to \varprojlim^1 H_{k+1}(X, M_i) \to H_k(X, \varprojlim M_i) \to \varprojlim H_k(X, M_i) \to 0$$

roof. Consider $\theta : \Pi M_i \to \Pi M_i$ given by $\theta\{m_i\} = \{m_i - f_i\ (m_{i+1})\}$. hen $\operatorname{Ker} \theta \cong \varprojlim M_i$ and coker $\theta \cong \varprojlim^1 M_i$. Hence, there is a short exact equence (*) : $0 \to \operatorname{Ker} \theta \to \Pi M_i \to \Pi M_i \to 0$. The exact sequence in question then ollows at once from the long exact homology sequence of (*) since, X being f finite type, $H_j(X,-)$ commutes with products.

emark. The condition $\varprojlim^1 M_i = 0$ is of course fulfilled if all the f_i are urjective or, more generally, if the system is Mittag-Leffler.

. Finite domination

Let X be a connected complex. We will write $cd(X,M)$ for $\sup\{i | H^i(X,M) \neq 0\}$ where M denotes a $\pi_1 X$-module (local coefficients) and, as ısual, we write $cd\ X$ for $\sup\{cd(X,M) | M$ is a $\pi_1 X$-module$\}$. Similarly, we lefine $hd\ X$ and $hd(X,M)$ by using homology (with local coefficients) instead of cohomology. (Hence, using the notation of [4], one has $d(X;M) = cd(X;M)$ and, if $M = \mathbb{Z}/p$ as trivial $\pi_1 X$-module, $d(X;\mathbb{Z}/p) = cd(X;\mathbb{Z}/p) = hd(X;\mathbb{Z}/p)$). Recall that for X of finite type one has $cd\ X = hd\ X$ (cf. [4, Theorem A 1]).

From Wall's work it is well known that X is finitely dominated if and only if X is of finite type and $cd\ X < \infty$ (cf. [7]). Let $\bar{X} \to X$ be a finite covering. Clearly, if X is finitely dominated then so is $\bar{X}$. Our main result provides reasonable conditions under which the converse holds.

Theorem 2.1. Let $\bar{X} \to X$ be a finite covering with $\bar{X}$ finitely dominated nilpotent. Then X is finitely dominated if and only if $hd(X,\mathbb{Z}) < \infty$ (i.e. $H_i X = 0$ for i sufficiently large). Furthermore, if X is finitely dominated then $hd(X,\mathbb{Z}) \leq hd\ X = hd\ \bar{X}$.

For the proof of the theorem we will first have to study the behavior of $cd(X)$ under passage to finite covering spaces. If $\overline{X} \to X$ is a finite covering then $cd\,\overline{X} \leq cd\,X$ and $cd\,\overline{X} = cd\,X$ in case $cd\,X < \infty$ (cf. [4]). The problem is then to find simple conditions ensuring $cd\,X < \infty$, assuming that $cd\,\overline{X} < \infty$.

Lemma 2.2. Let X be a space with fundamental group $\pi_1 X = \pi$ and let $\overline{\pi} \subset \pi$ be a normal subgroup with $G = \pi/\overline{\pi}$ a finite p-group. Suppose M is a π-module, which is nilpotent as a $\overline{\pi}$-module. Then

$$hd(X, M \otimes \mathbb{Z}/p) \leq hd(X, \mathbb{Z}/p)$$

Proof. Let $\overline{I}$ denote the augmentation ideal of $\mathbb{Z}\overline{\pi}$. Then $\overline{I}M$ is a $\pi_1 X$-submodule of M and, writing N for $M \otimes \mathbb{Z}/p$ and noticing that $\overline{I}^k N/\overline{I}^{k+1}N$ is trivial as $\overline{\pi}$-module, we infer $hd(X, \overline{I}^k N/\overline{I}^{k+1}N) \leq hd(X;\mathbb{Z}/p)$ by Lemma 4.1 of [4]. Since $\overline{I}^k N = 0$ for k large, this implies $hd(X, M \otimes \mathbb{Z}/p) \leq hd(X, \mathbb{Z}/p)$, as one can see from the obvious long exact homology sequences.

Notice that $hd(X, M \otimes \mathbb{Z}/p) \leq n$ implies that $H_i(X,M)$ is uniquely p-divisible for $i > n$. Hence $hd(X,M) \leq n$, if $hd(X, M \otimes \mathbb{Z}/p) \leq n$ for all primes p and $hd(X, M \otimes Q) \leq n$.

Theorem 2.3. Let $\overline{X} \to X$ denote a finite regular covering of degree a power of the prime p. Suppose $\overline{X}$ is a nilpotent space of finite type. Then for every $\pi_1 X$-module M one has

$$hd(X, M \otimes \mathbb{Z}/p) \leq hd(X, \mathbb{Z}/p)$$

Proof. We may assume that M is finitely generated, since for $M = \bigcup M_\alpha$ one has $hd(X, M \otimes \mathbb{Z}/p) \leq \sup\{hd(X, M_\alpha \otimes \mathbb{Z}/p)\}$. Write N for $M \otimes \mathbb{Z}/p$. Then, by [3, Proposition 4], one has

$$H_i(\overline{X}, N) \cong H_i(\overline{X}, \varprojlim N/\overline{I}^k N)$$

since $\overline{X}$ is nilpotent with $\pi_1 \overline{X}$ finitely generated and since N is finitely generated ($\overline{I}$ denotes the augmentation ideal of $\mathbb{Z}\pi_1\overline{X}$). Therefore we conclude from the covering spectral sequence

$$H_i(\pi_1 X/\pi_1 \overline{X}; H_j(\overline{X},-)) \Rightarrow H_{i+j}(X,-)$$

at

$$H_i(X,N) \cong H_i(X, \varprojlim N/\bar{I}^k N).$$

otice that since $N/\bar{I}^k N$ is a finitely generated nilpotent $\mathbb{Z}/p[\pi_1\overline{X}]$-module, the derlying abelian group is finite and hence $H_j(\overline{X}, N/\bar{I}^k N)$ is finite for all $j > 0$. rom the covering spectral sequence it is then obvious that $H_j(X,N/\bar{I}^k N)$ is finite r $j > 0$. Hence the inverse system $\{H_j(X,N/\bar{I}^k N)\}$ is Mittag-Leffler and we onclude from Lemma 1.4 that

$$H_j(X,N) \cong \varprojlim H_j(X,N/\bar{I}^k N).$$

rom Lemma 2.2 we know that $hd(X,N/\bar{I}^k N) \leq hd(X,\mathbb{Z}/p)$. Therefore $hd(X,N) \leq$ $d(X,\mathbb{Z}/p)$.

heorem 2.4. Let $\overline{X} \to X$ be a finite covering with $\overline{X}$ finitely dominated ilpotent. Assume that $hd(X,\mathbb{Z}) < \infty$. Then

$$hd(\overline{X}) = hd(X).$$

roof. First we consider the case where $\overline{X} \to X$ is a regular covering. Let $= \pi_1 X/\pi_1\overline{X}$ and factor $\overline{X} \to X$ as $\overline{X} \to X(p) \to X$ where $\overline{X} \to X(p)$ is a regular overing with $\pi_1 X(p)/\pi_1\overline{X}$ a p-Sylow subgroup of G. Since $X(p) \to X$ is a overing with degree prime to p, the transfer $H_i(X,M \otimes \mathbb{Z}/p) \to H_i(X(p),M \otimes \mathbb{Z}/p)$ s injective. Hence

$$hd(X,M \otimes \mathbb{Z}/p) \leq hd(X(p),M \otimes \mathbb{Z}/p) \leq hd(X(p),\mathbb{Z}/p)$$

here the second inequality follows from the previous theorem. Since X is initely dominated, $H^*(X;\mathbb{Z}/p)$ is finitely generated as abelian group. Since X s of finite type and $hd(X,\mathbb{Z}) < \infty$, $H^*(X;\mathbb{Z}/p)$ is a finitely generated abelian roup too. By a result of Quillen [6, Corollary 2.3] it follows then that $H^*(X(p);\mathbb{Z}/p)$ is finitely generated as module over the ring $H^*(X;\mathbb{Z}/p)$. Hence $d(X(p),\mathbb{Z}/p) < \infty$. Since $X(p) = \overline{X}$ for almost all primes p and since trivially $d(X,M \otimes \mathbb{Q}) \leq hd\ \overline{X}$, we conclude that $hd(X,M) < \infty$ for all M in an uniform ay, or $hd(X) < \infty$. It follows then that $hd\ X = hd\ \overline{X}$ (cf. [4], Proposition 1.1).

In case $\overline{X} \to X$ is not a regular covering, we choose a finite covering $Y \to X$ which is regular and which factors as $Y \to \overline{X} \to X$. Then $Y \to \overline{X}$ will be a finite regular covering with $\overline{X}$ finitely dominated nilpotent and hence Y is finitely dominated nilpotent. Hence $hd(\overline{X}) = hd(Y)$ and $hd(X) = hd(Y)$ by applying twice our result on regular coverings. Hence $hd\ \overline{X} = hd\ X$.

An example. If one applies Theorem 2.1 to an Eilenberg-MacLane space $X = K(G,1)$, then one obtains the following.

Corollary 2.5. Let G denote a group which possesses a finitely generated torsion-free nilpotent subgroup N of finite index. Suppose $H_i(G,\mathbb{Z}) = 0$ for i sufficiently large. Then $cd\ G = cd\ N < \infty$.

In particular, such a group G has to be a Poincaré duality group (in the sense of Bieri-Eckmann) of dimension $n = cd\ G = cd\ N$ (cf. [1, Theorem 3.3]).

References

[1] R. Bieri and B. Eckmann, Groups with homological duality generalizing Poincaré duality. Inventions Math. 20 (1973), 103-124.

[2] K. S. Brown, Homological criteria for finiteness. Comment. Math. Helv. 50 (1975), 129-135.

[3] K. S. Brown and E. Dror, The Artin-Rees property and homology. Israel J. Math. 22 (1975), 93-117.

[4] K. S. Brown and P. J. Kahn, Homotopy dimension and simple cohomological dimension of spaces. Comment. Math. Helv. 52(1977), 111-127.

[5] G. Mislin, Wall's obstruction for nilpotent spaces. Ann. of Math. 103 (1976), 547-556.

[6] D. Quillen, The spectrum of an equivariant cohomology ring I. Ann. of Math. 94 (1971), 549-572.

[7] C.T.C. Wall, Finiteness conditions for CW-complexes. Ann. of Math. 81 (1965), 56-69.

AN INTRODUCTION TO SHAPE THEORY

Jack Segal[(1)]

. Basic Ideas

In 1968 K. Borsuk [B_1] introduced the theory of shape which was a classifi-ation of compact metric spaces that was coarser than homotopy type but which oincided with it on absolute neighbourhood retracts (ANR's). His idea was to ake into account the global properties of compact metric spaces and neglect the ocal ones. Shape can be thought of as a sort of Čech homotopy type and its elationship to homotopy type is analogous to the relationship between Čech omology and singular homology.

Consider the following example. Let X denote the 1-sphere S^1 and let Y lenote the Polish circle, i.e., the union of the closure of the graph of $y = \sin \frac{1}{\pi}$, $0 < x \leq \frac{1}{\pi}$, and an arc from $(0,-1)$ to $(\frac{1}{\pi},0)$ which is disjoint from the graph except at its end points. Then X and Y are of different omotopy type but will turn out to be of the same shape. These spaces fail to be of the same homotopy type because there are not enough maps (continuous functions) of X into Y due to the failure of Y to be locally connected. Since any continuous image of X must be a locally connected continuum, it must be an arc in Y and so any such map is homotopically trivial. In other words, local difficulties prevent X and Y from being of the same homotopy even though globally they are very much alike (e.g., they both divide the plane into two components). Borsuk remedied this difficulty by introducing the notion of funda-mental sequence which is more general than that of mapping. For two compact subsets X and Y of the Hilbert cube I^∞ a fundamental sequence $\{f_n\} : X \to Y$ is a sequence of maps $f_n : I^\infty \to I^\infty$, such that for every neighbourhood V of Y there exists a neighborhood U of X and an integer n_0 such that for n, $m \geq n_0$

---

(1) Presented as a series of three lectures at the Algebraic Topology Conference, Vancouver, B.C., August 1977.

$$f_n|U \simeq f_m|U \text{ in } V.$$

Note that X is mapped into neighborhoods of Y but not necessarily Y itself and these neighborhoods possess nice local properties. Fundamental sequences are composed coordinatewise. Two fundamental sequences $\{f_n\}$, $\{g_n\} : X \to Y$ are said to be homotopic provided for every neighborhood V of Y there is a neighborhood U of X and in integer n_0 such that for $n \geq n_0$,

$$f_n|U \simeq g_n|U \text{ in } V.$$

The relation $\simeq$ is an equivalence on fundamental sequences. Two compacta X, Y in the Hilbert cube are said to be of the same shape, $\text{Sh}\, X = \text{Sh}\, Y$, if there are fundamental sequences $\{f_n\} : X \to Y$, $\{g_n\} : Y \to X$ such that

$$\{g_n\}\{f_n\} \simeq \{1_X\} \text{ and } \{f_n\}\{g_n\} \simeq \{1_Y\}$$

where $\{1_X\}$ indicates the identity fundamental sequence determined by the identity map $1_X : X \to X$. If X and Y have the same homotopy type, then $\text{Sh}\, X = \text{Sh}\, Y$. Borsuk also showed that compact ANR's of the same shape have the same homotopy type. As one can see, Borsuk's approach is very close to the geometric situation.

In 1970 S. Mardešić and the author [M-S$_2$] developed shape theory on the basis of inverse systems of ANR's. In this approach shapes are defined for arbitrary Hausdorff compacta. Maps between such systems are defined as well as a notion of homotopy of such maps. This homotopy relation classifies maps between ANR-systems and these classes are called shape maps.

Since any metric continuum can be represented as an inverse limit of an inverse sequence of ANR's (actually polyhedra [M-S$_1$]) in the metric case one can use ANR-sequences instead of ANR-systems. Compact metric spaces and shape maps form the shape category. Mardešić [M$_1$] generalized shape theory to arbitrary topological spaces. There is a functor from the category of metric spaces to the shape category which keeps spaces fixed and sends every map f into the shape map whose representative is any map $\underline{f}$ of ANR-sequences associated with f. (Note: while in the homotopy category every morphism has a representative which is a map, this is not true in the shape category.) The ANR-system approach yields a connuous theory, i.e., the shape functor commutes with taking inverse limits just as n the case of Čech homology. This is true for a single compactum or pairs of ompacta. Mardešić has shown that Borsuk's shape theory is not continuous on airs of compacta. So while the two approaches agree on compact metric spaces, hey differ on <u>pairs</u> of compact metric spaces. Borsuk's theory is the more geo- etrical of the two theories while the ANR-system approach is more categorical.

In addition to being more categorical, the ANR-system approach is useful in tudying the shape of a space X because any ANR-system expansion of X can be sed. In many cases the space X itself is defined by means of such a sequence e.g., solenoids are defined by an inverse sequence of circles) or can be btained as an inverse limit of an inverse sequence of nice spaces (e.g., manifold like spaces). This method has led to the shape classification of all (m-sphere)- ike continua [M-S_2] and (projective m-space)-like continua [M-S_1], and (complex rojective m-space)-like continua [W].

Two important shape invariants are Čech homology and cohomology [M-S_2]. In ddition, Borsuk [B_2] has introduced an interesting shape invariant called novability. Mardešić and Segal [M-S_3] have redefined movability in terms of ANR- systems. Movability can be defined in any pro-category and is useful in listinguishing between spaces when the standard invariants of algebraic topology fail to do so.

R. H. Fox's use of shape theory [F] to obtain a generalization of the fundamental theorem of covering spaces to the non-locally connected case illustrates shape theory's ability to eliminate local conditions and at the same time obtain a more general result. Many theorems of algebraic topology have a shape version which is easier to state and more intuitively satisfying.

T. A. Chapman [C] has obtained the following elegant characterization of the shape of metric compacta: Let X and Y be two metric compacta contained in the pseudo-interior of the Hilbert cube I. Then X and Y have the same shape if $I^\infty - X$ and $I^\infty - Y$ are homeomorphic. This result looks less surprising if one recalls that in ∞-dimensional manifolds homotopy and homeomorphism problems are often equivalent. Chapman's methods are those of ∞-dimensional manifold theory.

This approach to shape has defied generalization to the nonmetric case since a point and a closed interval can be embedded in the pseudo-interior of the uncountable product of closed intervals so that their complements are not homeomorphic.

In 1973 Mardešić $[M_1]$ described the shape category for topological spaces. This approach is much more categorical than Borsuk's and is based on the notion of shape map. The description which follows is due to G. Kozlowski who developed his version independently of Mardešić, however, the two theories are essentially the same (see $[K\text{-}S_2]$).

Let W be the category of all spaces having the homotopy type of a CW-complex and homotopy classes of maps between them. If X is a topological space, then Π_X is the functor from W to the category of sets and functions which assigns to a $P \in Ob(W)$ the set $\Pi_X(P) = [X,P]$ of all homotopy classes of maps of X into P and which assigns to any homotopy class $\phi: P \to Q$ between P, $Q \in Ob(W)$ the induced function $\phi_{\#}: [X,P] \to [X,Q]$ which maps the homotopy class $f{:}X \to P$ into the composition $\phi f = \phi_{\#}(f)$ of the homotopy classes of f and ϕ. A natural transformation Ψ of the functor Π_X into the functor Π_Y assigns to each homotopy class $f : X \to P$ a homotopy class $\Psi(f) : Y \to P$ in such a way that for all homotopy classes $f: X \to P$, $g: X \to Q$, and $\phi: P \to Q$ such that $\phi f = g$ we have $\phi\Psi(f) = \Psi(g)$. If $f{:}X \to Y$ is a map, then there is a natural transformation $f^{\#}: \Pi_Y \to \Pi_X$ which assigns to the homotopy class $\phi: Y \to P$ the composition $\phi[f] = f^{\#}(\phi)$ of the homotopy class $[f]$ of f with ϕ. A natural transformation from Π_Y to Π_X will be called a shape map from X to Y.

Given two spaces X and Y we say that X <u>shape dominates</u> Y if and only if there are natural transformations $\Phi{:}\Pi_Y \to \Pi_X$ and $\Psi{:}\ \Pi_X \to \Pi_Y$ such that $\Psi\Phi = 1^{\#}_Y$. If, in addition, $\Phi\Psi = 1^{\#}_X$, then X and Y are said to be of <u>the same shape</u>. In other words, X and Y have the same shape if and only if there is an invertible natural transformation (i.e., a natural equivalence) of the functors Π_X and Π_Y.

In place of W one could use any homotopy equivalent category (i.e., each space from one has the homotopy type of some space from the other) and get the same shape classification. Included among such categories are (possibly infinite) polyhedra, ANR's (metric), simplicial CW-spaces, simplicial CW-spaces with the metric topology.

In 1975 K. Morita [Mor] observed that the notion of a shape map of topological spaces can also be described using the ANR-systems approach of [M-S$_2$]. We follow the description given in [M$_2$]. Let K be an arbitrary category and associate with K a new category pro K whose objects are all inverse systems $\underline{X} = \{X_\alpha, p_{\alpha\alpha'}, A\}$ in K over all directed sets $(A;\leq)$. A map of systems $\underline{X} \to \underline{Y} = \{Y_\beta, q_{\beta\beta'}, B\}$ consists of a function $f\colon B \to A$ and of a collection of morphisms $\{f_\beta\}\colon X_{f(\beta)} \to Y_\beta$, $\beta \in B$, in K such that for $\beta \leq \beta'$ there is an $\alpha \geq f(\beta), f(\beta')$ such that $f_\beta p_{f(\beta)\alpha} = q_{\beta\beta'} f_{\beta'} p_{f(\beta')\alpha}$, or diagramatically,

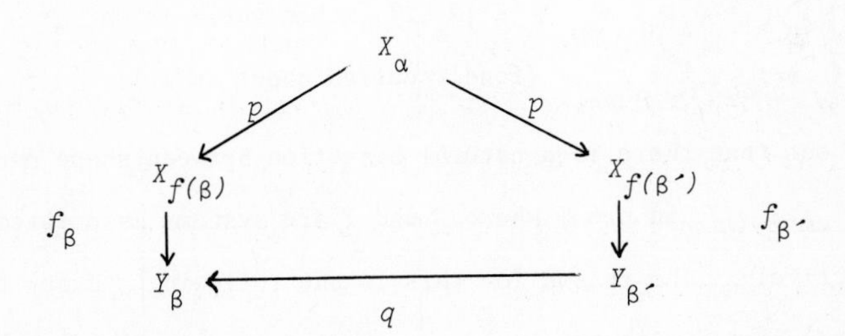

where we have deleted the subscript indexing from the bonding maps.

Two maps of systems $(f;\{f_\beta\}), (g;\{g_\beta\})\colon \underline{X} \to \underline{Y}$ are said to be equivalent provided for each $\beta \in B$ there is an $\alpha \geq f(\beta), f(\beta')$ such that $f_\beta p_{f(\beta)\alpha} = g_\beta p_{g(\beta)\alpha}$, or diagramatically,

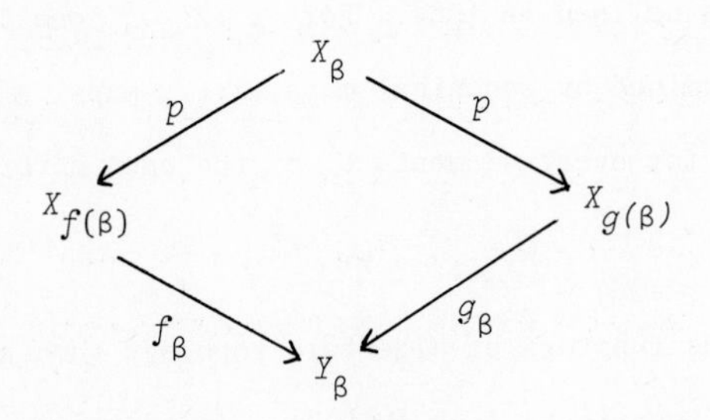

Morphisms $\underline{f}:\underline{X} \to \underline{Y}$ in pro K are equivalence classes of maps of systems $(f; \{f_\beta\}):\underline{X} \to \underline{Y}$. If $\underline{g}:\underline{Y} \to \underline{Z} = \{Z_\gamma, r_{\gamma\gamma'}, C\}$ is given by $(g;\{g_\gamma\})$, then the composition $\underline{g}\,\underline{f}:\underline{X} \to \underline{Z}$ is given by $(fg;\{g_\gamma f_{g(\gamma)}\})$. The identity $\underline{1}_{\underline{X}}:\underline{X} \to \underline{X}$ is given by $(1_A;\{1_{X_\alpha}\})$.

Morita generalized the ANR-system approach of [M-S_2] by calling an inverse system $\underline{X} = \{X_\alpha, p_{\alpha\alpha'}, A\}$ in W associated with a topological space X provided there exist homotopy classes of maps $p_\alpha : X \to X_\alpha$ such that implies $p_\alpha = p_{\alpha\alpha'} p_{\alpha'}$, i.e., the usual composition of projections and bonding maps and the following two conditions hold for every $P \in Ob(W)$:

(1) for every homotopy class of maps $\phi \in [X,P]$ there is an $\alpha \in A$ and a homotopy class of maps $\phi_\alpha \in [X_\alpha,P]$ such that $\phi = \phi_\alpha p_\alpha$ (i.e., each ϕ factors through some X_α)

and

(2) whenever $\phi_\alpha p_\alpha = \psi_\alpha p_\alpha$, ϕ_α, $\psi_\alpha \in [X_\alpha,P]$, then there is an $\alpha' \geq \alpha$ such that $\phi_\alpha p_{\alpha\alpha'} = \psi_\alpha p_{\alpha\alpha'}$ (fondly called short tails).

Morita points out that there is a natural bijection between shape maps $X \to Y$ and morphisms $\underline{f}:\underline{X} \to \underline{Y}$ in pro W, where $\underline{X}$ and $\underline{Y}$ are systems associated with X and Y respectively. The reason for this is the following. Every topological space X is associated with the inverse system $\underline{X}$ in W formed by the nerves of all open locally-finite normal coverings of X. An open covering U of X is normal provided there exists a sequence of open coverings U_n such that $U_0 = U$ and U_{n+1} is a star-refinement of U_n. The existence of canonical mapping shows that open locally-finite normal coverings coincide with open locally-finite numerable coverings as defined in [Do]. For $p_\alpha : X \to Y$ one takes (unique) homotopy classes determined by canonical maps, i.e., maps $\phi_\alpha : X \to Y$ such that $(\phi_\alpha)^{-1}(\mathrm{St}(U,X_\alpha)) \subset U$ for every element U of the open covering U.

2. Shape Invariants

Various continuous functors of algebraic topology such as Čech homology (or cohomology) are shape invariants (see [M-S_2]). In addition, it is possible to describe new continuous functors for an arbitrary topological space X such as the shape groups by taking inverse limits of inverse systems of homotopy groups of inverse systems associated with (see [M_2]). Furthermore, if one does not pass to the limit in this situation, one obtains the homotopy pro groups which are a more delicate shape invariant. We describe these groups and pro groups now following [M_2].

Let W_0 denote the category of all pointed spaces having the homotopy type of a pointed CW-complex and pointed homotopy classes of maps between them. For $k \geq 1$, define the k-th homotopy pro group of (X,x) in W_0 as

$$\Pi_k(\underline{X},\underline{x}) = \{\Pi_k(X_\alpha,x_\alpha),\ p_{\alpha\alpha'\#},A\}$$

where $p_{\alpha\alpha'\#}$ also depends on k but we supress it notationally and $(\underline{X},\underline{x}) = \{(X_\alpha,x_\alpha),p_{\alpha\alpha'},A\}$ is an inverse system in W_0 associated with (X,x). Every map of systems $\underline{f}:(\underline{X},\underline{x}) \to (\underline{Y},\underline{y})$ determines a homomorphism of pro groups

$$\underline{f}_\#:\Pi_k(\underline{X},\underline{x}) \to \Pi_k(\underline{Y},\underline{y})$$

in a functorial way so that homotopy equivalent systems in W_0 have isomorphic homotopy pro groups. The k-th shape group of (X,x) is defined as $\check{\Pi}_k(X,x) = \varprojlim \Pi_k(\underline{X},\underline{x})$ where $(\underline{X},\underline{x})$ is an inverse system associated with (X,x).

In general, the homotopy pro groups carry more information than the shape groups. The next shape invariant, called movability, was originally introduced by Borsuk [B_2] for metric compacta. It is a far reaching generalization of the notion of ANR. Borsuk's description was very geometric. Mardešić and Segal generalized this notion to compacta using the ANR-system approach in [$M\text{-}S_2$]. In [Mos_1] Moszynska defined an apparently stronger property called uniform movability for compacta which turned out to be the same in the metric case. However, Kozlowski and Segal in [$K\text{-}S_1$] gave a categorical description of this property which applied to arbitrary topological spaces and showed that movability was stronger than movability on compacta. The importance of this notion in shape theory stems from the fact that in its presence one may take inverse limits without losing information. Moreover, to generalize various classical theorems of algebraic topology to the non-metric or non-compact case one needs the full strength of uniform movability.

An inverse system $\underline{X} = \{X_\alpha, p_{\alpha\alpha'}, A\}$ in W is said to be uniformly movable provided:

(1) it is movable, i.e., for every $\alpha \in A$, there exists an $\alpha' \in A$, $\alpha' \geq \alpha$ such that for all $\alpha'' \in A$, $\alpha'' \geq \alpha$, there exists a map $r^{\alpha'\alpha''} : X_{\alpha'} \to X_{\alpha''}$ such that

$$p_{\alpha\alpha''} r^{\alpha'\alpha''} \simeq p_{\alpha\alpha'}$$

and

(2) the r's form a map of systems, i.e., $\{r^{\alpha'\alpha''}\}X_{\alpha'} \to \{X_{\alpha''}, p_{\alpha''\alpha'''}, \alpha'' \geq \alpha\}$. Then a space X is said to be uniformly movable if it has an associated uniformly movable inverse system.

For example, the solenoids are not movable. Let $Q = (q_1, q_2, \ldots)$ be a sequence of primes. Denote by $\underline{S}^1_Q$ the inverse system $\{X_n, p_{nn+1}\}$ with $X_n = S^1$ and $p_{nn+1}: X_{n+1} \to X_n$ being a map of degree q_n. Then the shape of the inverse limit S_Q^1 of the inverse sequence $\underline{S}_Q^1$ is completely determined. Suppose S_Q^1 is movable. We show that this assumption leads to a contradiction. By definition there is an $n' \geq 1$ such that for each $n'' \geq n$ there exists a map $r^{n'n''} : X_{n'} \to X_{n''}$ such that the following diagram commutes up to homotopy

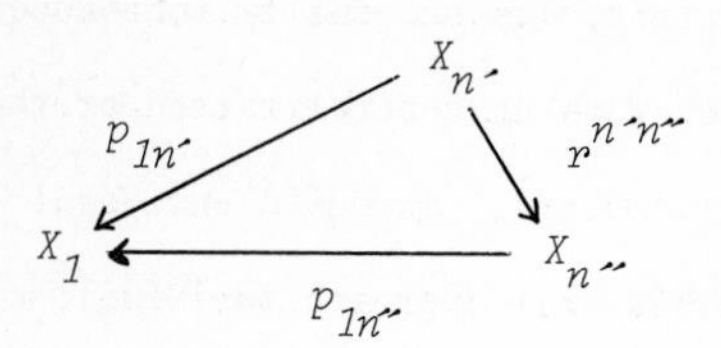

i.e., $p_{1n'} \simeq p_{1n''} r^{n'n''}$.

Taking degrees we have

$$\deg p_{1n'} = \deg(p_{1n''} r^{n'n''}) = \deg p_{1n''} \cdot \deg r^{n'n''}$$

i.e.,

$$q_1 \cdot q_2 \cdots q_{n'-1} = q_1 \cdot q_2 \cdots q_{n''-1} \cdot \deg r^{n'n''} .$$

hen dividing by $q_1 \cdots q_{n-1}$ we get

$$1 = q_{n'} \cdots q_{n''-1} \cdot \deg r^{n'n''}$$

which is impossible since the q's are primes. Notice also that in this example (non-movable) one loses information by taking the inverse limit of the first homology pro group $H_1(\underline{X}) = \{H_1(X_n), p_{nn+1*}\}$ where $H_1(X_n) \approx Z$ and $p_{nn+1*} : Z \to Z$ is multiplication by q_n. The first Čech homology group $\check{H}_1(X) = \varprojlim H_1(\underline{X}) = 0$, but $H_1(\underline{X})$ is not the zero pro group.

Following the more categorical approach of [K-S$_1$] we can also say a topological space X is uniformly movable provided, that for each map $f:X \to P$ of X into a (possibly infinite) polyhedron P, there exists a polyhedron Q and natural transformations $\Phi:\Pi_X \to \Pi_Q$, $\Psi:\Pi_Q \to \Pi_X$ such that

$$\Psi\Phi[f] = [f].$$

Since any natural transformation $\Psi:\Pi_Q \to \Pi_X$ satisfies $\Psi = g^{\#}$ for a map (unique up to homotopy) $g:X \to Q$, the above condition can be stated: for each map $f:X \to P$ there exists a polyhedron Q, maps $g:X \to Q$, $\phi:Q \to P$, and a natural transformation $\Phi:\Pi_X \to \Pi_Q$ such that $\phi g \simeq f$ and $\Phi[f] = [\phi]$. Furthermore, Kozlowski and Segal show that if (X,x) is a uniformly movable pointed continuum with shape groups $\check{\Pi}_n(X,x) = 0$, for all $n \geq 1$, then (X,x) has the shape of a point. It follows from this that the only uniformly movable compact connected abelian topological group with $\check{\Pi}_1(X,x) = 0$ is the trivial group. This failure is a result of the fact that X is not movable. The proof that movability is a shape invariant actually carries over to a more general categorical setting. In fact, one can define movability in pro K for any category K. Furthermore, the proof that movability is preserved under shape domination implies that a covariant functor from one category to another carries movable systems in the first category to movable systems in the second. So if $\underline{X}$ is a movable system in W_0 then $H_k(\underline{X})$ are movable pro groups for $k \geq 1$.

It is interesting to note that each movable pro group $\underline{G} = \{G_\alpha, h_{\alpha\alpha'}, A\}$ has the Mittag-Leffler property, i.e., for each $\alpha \in A$ there is a $\beta \geq \alpha$ such that for each $\gamma \geq \beta$, $h_{\alpha\beta}(G_\beta) = h_{\alpha\gamma}(G_\gamma)$. This in turn implies that the first derived functor is trivial, i.e., $\lim^1 \underline{G} = *$ and since for a sequence of abelian groups $\underline{G}$ set-theoretic reasons imply $\varprojlim^k \underline{G} = *$ for $k \geq 2$ there is no loss of information in taking $\varprojlim$ of $\underline{G}$.

An extensive literature exists which relates algebraic properties of a discrete group G to topological properties of a topological group, namely, its character group char G (see [P]). For example, if G is discrete, then char G is compact; if G is discrete and torsion free, then char G is compact and connected. Moreover, by the use of Pontryagin duality the converse is also true. It is also shown in [P] that G has property L iff char G is locally connected. N. Steenrod [S] showed that for a compact connected abelian topological group X, $\check{H}^1(X) \approx$ char X. So for such groups if Sh X = Sh Y their first Čech cohomology groups are isomorphic. Therefore char $X \approx$ char Y and so $X \approx Y$ (as topological groups). J. Keesling [K_1], [K_2], [K_3], [K_4] has made an extensive study of the uses of shape theory on topological groups. For example, he showed for a compact connected abelian topological group X movability was equivalent to the char X having property L. An abelian group G is said to have property L, if every finite subset of G is contained in a finitely generated subgroup that admits division. A subgroup H of G admits division, if whenever $g \in G$ and n is a positive integer, then $ng \in H$ implies $g \in H$. This equivalent to saying G/H is torsion free. This terminology arose in the classical theory of topological groups. More modern usage says an abelian group G is $\aleph_1$-free if every countable subgroup H of G is free abelian. A famous theorem of Pontryagin (namely, a torsion free abelian group G is $\aleph_1$-free iff every subgroup of finite rank is free) implies that a torsion free abelian group G is $\aleph_1$-free iff G has property L.

A crucial fact in using shape theory on topological groups is that each homotopy class of maps $\phi: X \to Y$ between arcwise connected topological groups contains a unique continuous homomorphism. From this it follows that for each shape map $\underline{f}: X \to Y$, there exists a unique continuous homomorphism h (such that $h^{\#} = \underline{f}$) which determines $\underline{f}$. Another useful classical result is that every compact connected abelian topological group X is the inverse limit of an inverse system of tori and continuous homomorphisms. Using these two facts it is shown in [K-S$_1$] that if a compact connected abelian topological group X is uniformly movable, then $(X,0)$ is uniformly movable. Making use of the toroidal inverse limit representation of X one gets $\check{\Pi}_n(X,0) = \varprojlim\{\Pi_n(X_\alpha,0), p_{\alpha\alpha'\#}\} = 0$ for $n \geq 2$. Hence if $\check{\Pi}_1(X,0) = 0$ we can make use of a special case of an infinite-dimensional shape version of the Whitehead theorem [K-S$_1$] to get that X is the zero group. This special case says if (X,x) is a uniformly movable pointed continuum with $\check{\Pi}_n(X,x) = 0$ for all n, then (X,x) has trivial shape. Finally, this information about topological groups is used to produce an example of a non-trivial compact connected abelian topological group X which is movable and $\check{\Pi}_1(X,0) = 0$ but is not uniformly movable. So while movability and uniform movability agree on compact metric spaces uniform movability is a stronger property on compact Hausdorff spaces.

3. Some Classical Theorems of Algebraic Topology in Shape Theory

One of the motivating ideas in the development of shape theory was that theorems in homotopy theory valid only for CW-complexes or spaces with strong local properties should be true in shape theory for arbitrary spaces with certain "corrections." Recall Whitehead's classical theorem: Let (X,x), (Y,y) be connected CW-complexes, $n_o = \max(1 + \dim X, \dim Y)$ and $f:(X,x) \to (Y,y)$ be a map such that the induced homomorphism

$$f_{k\#}: \Pi_k(X,x) \to \Pi_k(Y,y)$$

is an isomorphism for $1 \leq k < n_o$ and is an epimorphism for $k = n_o$, then f is a homotopy equivalence. The importance of this theorem lies in the fact that it translates strictly algebraic information into homotopy information. Now the following shape version of the Whitehead theorem has been developed in successively more generality by Moszyńska [Mos$_2$], Mardešić [M$_2$] and Morita [Mor].

<u>The Whitehead theorem in shape theory</u>: Let (X,x), (Y,y) be connected topological spaces, $n_o = \max(1 + \dim X, \dim Y) < \infty$ and $\underline{f}:(X,x) \to (Y,y)$ be a shape map such that the induced homomorphism

$$f_{k\#}:\Pi_k(\underline{X},\underline{x}) \to \Pi_k(\underline{Y},\underline{y})$$

is an isomorphism of pro groups for $1 \leq k < n_o$ and an epimorphism for $k = n_o$, then $\underline{f}$ is a shape equivalence.

Note that the "correction" required here is replacing the homotopy groups by the homotopy pro groups, Mardešić showed that the proof of the shape version of the Whitehead theorem reduces to a shape version of the Fox theorem by considering the "mapping cylinder" of $\underline{f}$ and by applying the exactness of the homotopy pro groups to the pair composed of this mapping cylinder and X.

One cannot do away with the dimension restriction in the shape version of the Whitehead theorem. Kahn [K_2] using Adam's work on K-theory constructed an ∞-dimensional continuum X which has trivial homotopy pro groups but is not of trivial shape (in fact, it is not movable). A map sending X to a one-point space shows the need for a dimension restriction in the shape version of the Whitehead theorem. There is, however, a movable version of this theorem for continua without a dimension restriction.

Now recall the Fundamental Theorem of Covering Spaces: The d-fold covering spaces of a connected, locally connected and semi locally 1-connected space X are in biunique correspondence with representation classes of $\Pi_1(X)$ in the symmetric group Σ_d. R. H. Fox [F] used shape theory to generalize the theory of covering spaces in non-locally well-behaved metrizable spaces. He described a shape version of covering spaces called overlays (which is just a literal translation of Überlagerung). There are open covers satisfying a certain intersection property designed to avoid a difficulty first pointed out by Zabrodsky. Not every covering is an overlay. But Fox shows that a covering $e:X \to Y$ is an overlaying if Y is locally connected or e has finite degree.

The following is a result on which most of the considerations which istinguish overlay theory from covering space theory are based. Extension heorem: If Y is a subset of a metrizable space Q, and $e:X \to Y$ is any overaying of Y, then e can be extended to an overlaying $f:U \to V$, where V is a suitable chosen neighbourhood of Y in Q and U is a suitable superspace of X. On the other hand a covering $e:X \to Y$ might not be extendable over any neighborhood of Y.

The role of the fundamental group $\Pi_1(X,x)$ in the fundamental theorem of covering space theory is replaced by the first homotopy pro group $\Pi_1(\underline{X})$. With this correction Fox obtained for overlays a Lifting Theorem: Let X, Y and Z be connected metrizable spaces, and let $e:(X,x_o) \to (Y,y_o)$ be an overlaying. Then a mapping $g:(Z,z_o) \to (Y,y_o)$ can be lifted to a mapping $f:(Z,z_o) \to (X,x_o)$ iff $g_*(\Pi_1(\underline{Z},\underline{Z}_o)) \subset e_*(\Pi_1(\underline{X},\underline{x}_o))$. Then Fox's theory culminates with the following most elegant theorem. Fundamental Theorem of Overlays: Let Y be any metrizable space. Then there is a biunique correspondence between the d-fold overlays $(X,x_o) \to (Y,y_o)$ and the representations of $\Pi_1(\underline{X},\underline{x}_o)$ in the symmetric group of degree d. In contradistinction to the Fundamental Theorem of Covering Spaces this theorem is freed from all local assumptions on the space.

Although as originally conceived shape theory was designed to deal with global properties of metric compacta, it is also related to the local properties of paracompacta. Kozlowski and Segal [K-S$_2$] showed that any LC^n paracompactum X of dimension $\leq n$ is shape dominated by a polyhedron of dim $\leq n$. From this and other results it follows that such an X is an ANSR (absolute neighborhood shape retract), thus providing a shape version of the classical result than a LC^n compact metric space of dimension is an ANR. The importance of the classical result lay in the fact that it translated a local homotopy property into information about extending maps. This shape version likewise yields information about extending shape maps.

In [K-S$_3$] it is shown using partial realization techniques that for LC^n paracompacta the shape groups and the homotopy groups are naturally isomorphic. An analogous result for Čech homology groups and singular homology groups had been obtained earlier by Mardešić [M_4].

In [K-S_2] the notion of extensor is generalized to shape theory for paracompacta. Likewise the notions of FANR [B_4] and ANSR [M_3] are generalized to paracompacta. The starting point is the generalization of the neighborhood extension of maps to the neighborhood extension of shape morphisms. The universal qualification of this property gives the concept of absolute neighborhood shape extensor (ANSE).

A space Y is said to be an absolute neighborhood shape extensor for paracompacta (ANSE) if for any natural transformation $\Phi:\Pi_Y \to \Pi_A$, where A is any closed subset of an arbitrary paracompactum X, there is a closed neighborhood N of A and a natural tranformation $\Psi:\Pi_Y \to \Pi_N$ such that $\rho\Psi = \Phi$ (where $\rho:\Pi_N \to \Pi_A$ denotes the restriction). In the ANR-systems approach this implies that any compactum Y is an ANSE if any shape map $\underline{f}:\underline{A} \to \underline{Y}$ can be extended to a shape map $\underline{F}$ of a closed neighborhood N of A in X. Here $\underline{F}$ extends $\underline{f}$ means $\underline{F}\,\underline{i} \simeq \underline{f}$ where $\underline{i}$ is a shape map of A into N induced by the inclusion $i:A \to N$.

A paracompactum Y is said to be an absolute neighborhood shape retract (ANSR) if, whenever Y is a closed subset of a paracompactum Z, there exist a neighborhood N of Y in Z and a natural transformation $\Psi:\Pi_Y \to \Pi_N$ such that $\rho\Psi = 1_{\Pi_Y}$. This generalizes the notion of ANSR due to Mardešić [M_3] to paracompacta in shape theory. Mardešić's definition was a generalization of Borsuk's [B_4] fundamental absolute neighborhood retracts (FANR's) to the compact Hausdorff case. Every compact ANSR is an ANSR (in the sense of Mardešić) since for any natural transformation $\Phi:\Pi_Y \to \Pi_X$ there exists a map of systems $\underline{f}:\underline{X} \to \underline{Y}$ such that $\underline{f}^{\#} = \Phi$. If a paracompactum Y is an ANSE, then it is an ANSR.

We now summarize how shape theory can be used effectively to deal with some local homotopy properties of paracompacta. We describe the results of [K-S_2], as well as, classical results on locally well-behaved compacta in diagram form. An arrow ($\to$) indicates class inclusion and a broken arrow ($\text{-}n\!\to$) indicates class inclusion under the additional hypothesis that the (covering) dimension of the pace in question is $\leq n$. Here SDP indicates a space which is shape dominated y a polyhedron.

Classically, we have for metric spaces:

$$LC^n \xleftarrow{\qquad} ANE \leftrightarrow ANR$$
$$\xrightarrow{\quad n \quad}$$

nd for compacta:

$$LC^n \leftarrow ANE \leftrightarrow ANR$$

ince an ANSR may behave badly locally there is no chance of extending the metric esult, $ANR \rightarrow LC^n$, to paracompacta. On the other hand, an example due to . W. Saalfrank shows that the metric result, at most n-dimensional and $C^n \rightarrow ANR$, cannot be extended to compacta. However, it does extend to paracompacta n shape theory, i.e., at most n-dimensional and $LC^n \rightarrow ANSR$.

Uniform n-movability is a stratification of uniform movability (see [K-S_2]). very LC^{n-1} paracompactum is uniformly n-movable. So every LC^{n-1} paracompactum f dimension $\leq n$ is uniformly movable. Then in shape theory we have for ompacta:

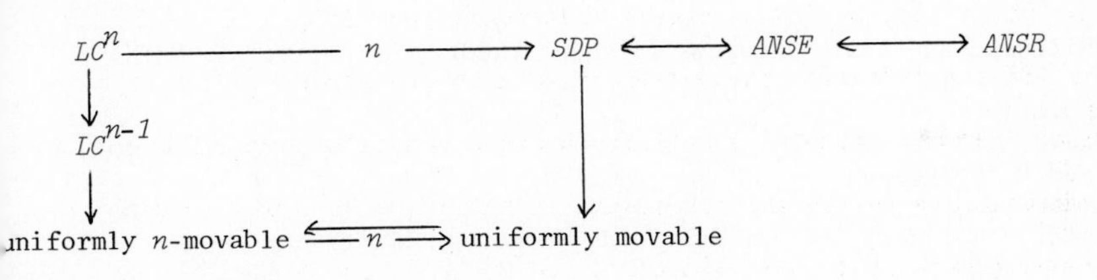

and for paracompacta

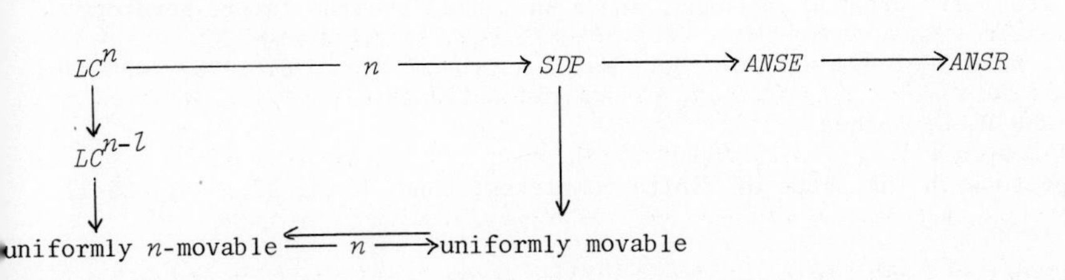

It is not known if $ANSR \rightarrow ANSE$ or if $ANSE \rightarrow SDP$ for paracompacta. The role of *SDP* for pointed compacta has been investigated extensively by D. A. Edwards and R. Geoghegan [E-G_1] and [E-G_2] and by R. Geoghegan and R. C. Lacher [G-L]. In [E-G_1] they obtain a shape version of the Wall obstruction.

REFERENCES

K. Borsuk

[B_1] Concerning homotopy properties of compacta, Fund. Math. 62(1968), 223-254.

[B_2] On movable compacta, Fund. Math. 66(1969), 137-146.

[B_3] Theory of shape, Monografie Matematyczne 59, Polish Science Publications, Warszawa, 1975.

[B_4] Fundamental retracts and extensions of fundamental sequences, Fund. Math. 64(1969), 55-85.

T. A. Chapman

[C] On some applications of infinite-dimensional manifolds to the theory of space, Fund. Math. 76(1972), 181-193.

A. Dold

[Do] Lectures on Algebraic Topology, Springer-Verlag, Berlin, 1972.

D. Dydak

[Dy] The Whitehead and Smale theorems in shape theory, Dissertations Math. (to appear).

D. A. Edwards and R. Geoghegan

[E-G_1] Shapes of complexes, ends of manifolds, homotopy limits and the Wall Obstruction, Ann. of Math. 101(1975), 521-535. Correction 104(1976), 389.

[E-G_2] The stability problem in shape, and a Whitehead theorem in pro-homotopy, Bull. Acad. Polon. Sci. Ser. Sci. Math. Astronom, Phys. 81(1975), 438-440.

R. Geoghegan and R. C. Lacher

[G-L] Compacta with the shape of finite complexes, Fund. Math. 92(1976), 25-27.

R. H. Fox

[F] On shape, Fund. Math. 74(1972), 47-71.

D. Handel and J. Segal

[H-S] Shape classification of (projective n-space)-like continua, Gen. Top. and its Appl. 3(1973), 111-119.

D. S. Kahn

[Ka] An example in Čech cohomology, Proc. Amer. Math. Soc. 16(1965), 584.

J. Keesling

[K_1] On the shape of torus-like continua and compact connected topological groups, Proc. Amer. Math. Soc. 40(1973), 297-302.

[K_2] Shape theory and compact connected abelian topological groups, Trans. Amer. Math. Soc. 194(1974), 349-358.

[K_3] An algebraic property of the Čech cohomology groups which prevents local connectivity and movability, Trans. Amer. Math. Soc. 190(1974), 151-162.

[K_4] The Čech cohomology of compact connected abelian topological groups with applications to shape theory, Lecture Notes in Math. 438, Berlin 1975, 325-331.

G. Kozlowski and J. Segal

[$K\text{-}S_1$] Movability and shape-connectivity, Fund. Math. 93(1976), 145-154.

[$K\text{-}S_2$] Locally well-behaved paracompacta in shape theory, Fund. Math. 95(1977), 55-71.

[$K\text{-}S_3$] Local behavior and the Vietoris and Whitehead theorems in shape theory, Fund. Math. (to appear).

S. Mardešić

[M_1] Shapes for topological spaces, Gen. Top. Appl. 3(1973), 265-282.

[M_2] On the Whitehead Theorem I, Fund. Math. 91(1976), 51-64.

[M_3] Retracts in shape theory, Glasnik Mat. Ser. III 6(26) (1971), 153-163.

[M^4] Comparison of singular and Čech homology in locally connected spaces, Michigan Math. J. 6(1959), 151-166.

S. Mardešić and J. Segal

[$M\text{-}S_1$] ε-mappings onto polyhedra, Trans. Amer. Math. Soc. 109(1963), 146-164.

[$M\text{-}S_2$] Shape of compacta and ANR-systems, Fund. Math. 72(1971), 41-59.

[M-S_3] Movable compacta and ANR-systems, Bull. Acad. Polon. Sci. Ser. Sci. Math. Astronom. Phys. 18(1970), 649-654.

K. Morita

[Mor] On shapes of topological spaces, Fund. Math. 86(1975), 251-259.

M. Moszyńska

[Mos_1] Uniformly movable compact spaces and their algebraic properties, Fund. Math. 77(1972), 125-144.

[Mos_2] The Whitehead theorem in the theory of shapes, Fund. Math. 80(1973), 221-263.

L. A. Pontryagin

[P] Topological Groups (2nd Edition), Gordon and Breach, New York, 1966.

N. Steenrod

[S] Universal homology groups, Amer. J. Math. 58(1936), 661-701.

T. Watanabe

[W] Shape classifications for complex projective space-like and wedges of n-sphere-like continua, Sci. Rep. of the Tokyo Kyoiku Daigaku, Sec. A, 12(1975), 233-245.

GENERATING CURTIS TABLES

BY

MARTIN C. TANGORA

By a "Curtis table" I mean a chart of the E_2 term of the unstable Adams spectral sequence as obtained from the lambda algebra ([3],[8]). This note is a discussion of the problem of calculating such a table.

1. Introduction

For each prime p, the lambda algebra provides an E_1 term for the stable or unstable Adams spectral sequence. All the rules of the game are explicitly laid down and our problem is to organize the work efficiently. This calls for some ingenuity because the algebra is non-commutative and has many generators, so that the calculation has a formidable growth rate.

In my oral report at Vancouver I began by discussing how to produce tables of differentials and relations in Λ in an efficient way. A written version of this material will appear elsewhere [6]. Here I wish to move on to the problem of calculating the Curtis table.

On its face, the problem appears to be to calculate $H^{**}(\Lambda(n),d(n))$ for all n and a satisfying range of bi-gradings. Here $\Lambda(n)$ denotes the sub-algebra corresponding to S^n and $d(n)$ denotes the restriction of the differential. We can already revise the problem: instead of Λ we can work in a smaller sub-algebra with almost the same homology, and we can devise methods which emphasize certain features of the homology and ignore others. Such ideas will be set out in Section 2.

Since the lambda algebra was developed by a group of mathematicians, many of the ideas for dealing with it are folklore and it is difficult to give credit to individuals for those ideas. Moreover many results remain unpublished. In addition to the usual references (listed at the end) I know of the following work. G.W. Whitehead has done extensive claculations at $p = 2$ for projective space as well as spheres. Hans Salomonsen wrote a machine program around 1967, when Curtis was at Aarhus, to generate a Curtis table for $p = 2$ and $t\text{-}s \leq 23$. William Hansen, a student of Mahowald, wrote a machine program for $p = 2$ and $t-s \leq 30$ [4], and Mahowald has continued to work in the area. In the terminology of this paper, Salomonsen used a full-cycle Curtis method, but the methods of Whitehead, Hansen and Mahowald are not Curtis methods, though they share many features.

Curtis tables appear in the cited literature, with brief explanations of how to use them, but not enough about how to calculate them. I am deeply indebted to Ed Curtis for taking the time at Stanford in August 1976 to coach me in his method, on the basis of which I have written algorithms, proved them correct and finite, and programmed them for machine calculation.

I hope later to publish an efficient algorithm with proofs. The present discussion may be regarded as an expository introduction to that work. It is organized as follows. In Section 2 we present several ideas (not new) for managing the calculation. An example is given in Section 3 along with some terminology. In the following three sections we explain how to incorporate these ideas into algorithms for generating the Curtis table, and how to show that these methods actually work. In the last section we summarize the current status of this effort.

For simplicity of exposition we will emphasize the case $p = 2$. However, we are working on odd primes as well, and will comment on the odd-prime case from time to time.

We generally suppress the lambdas from the notation and write, for example, 10.1 + 83 for $\lambda_{10}\lambda_1 + \lambda_8\lambda_3$.

2. Ideas

In this section we present several ideas for simplifying or streamlining the calculation of a Curtis table. All these ideas are more or less folklore and I do not know how to attribute them to individuals, though certainly Ed Curtis has been their leading promoter. All were passed to me by Curtis in 1976, more or less explicitly, and most were implicit in the table calculated in 1967 by Salomonsen [3].

2.1 Always represent elements of the lambda algebra in terms of the allowable basis. This masks certain important relations but is well adapted to our purpose and is implicit in the Curtis table.

2.2. Instead of Λ work with Λ', the sub-algebra with additive basis consisting, in the case $p = 2$, of all monomials $\lambda_1\lambda_2\ldots\lambda_s$ such that s is odd. It is well known [7] that the only difference this makes in the homology is the loss of the infinite towers corresponding to $\Pi_n S^n$ and the infinite cyclic summand of $\Pi_{4n-1}S^{2n}$.

When p is odd, the analogue of Λ' is the sub-algebra generated by the lambdas only, without the mu's. Unfortunately this does not have the right homology at all; it corresponds to the algebra of reduced powers. It may well be, however, that the best approach to $H(\Lambda)$ for odd p is first to find the homology of this sub-algebra and then to proceed by other methods.

2.3. Order everything. Let λ_m be larger than λ_n if and only if $m > n$. Order monomials lexicographically. Write polynomials with their terms in decreasing order, and order the polynomials lexicographically by their terms. Similarly order sets of polynomials. In this way we determine a canonical representative for every homology class, a canonical basis for every group of cycles or boundaries, etc.

In any long routine calculation it is a good idea to have canonical representatives (cf. [5]), but the ordering is particularly effective with Λ because it is so perfectly responsive to the topological purpose: the maximal term of the minimal representative of a homology class contains the information about Hopf invariant and sphere of origin.

An algorithm for finding the minimal basis will be sketched in Section 4.

2.4. Display the work in terms of the Adams coordinates $(r,s) = (t-s,s)$ where r is the homotopy dimension, s is the Adams filtration, and $t = r+s$ is the internal degree. Then the Curtis table is produced in the format of the Adams spectral sequence as usually found in nature.

Perhaps this is not so much an idea as the absence of an idea. The method of Hansen and Mahowald departs here from our line, and emphasizes sphere of origin rather than Adams filtration.

2.5. In calculating a certain bi-grading (r,s) make full use of prior calculations, "prior" meaning smaller r and s. In other words, start in low dimensions and work up.

It is not important whether we proceed by row, column, or diagonal (corresponding to successively larger values of s, r, or t respectively.) The point is that the search for cycles at (r,s) is greatly facilitated by the full knowledge of cycles in lower dimensions, because of the following proposition [7]: If the cycle x is written in canonical form as $\lambda_n x_1 + x'$, then x_1 must itself be a cycle. This fact lies at the heart of the method, and connects with the EHP sequence: x_1 is the Hopf invariant of x, and the sphere of origin of x is n+1 [3].

The efficiency which this idea brings to the search for cycles is so great that it may become worth while to make considerable prior calculations even when only one bigrading is sought. This was the case when I needed the canonical form of the generator at (52,5) [6].

2.6. Suppress "invisible elements" not only from the table but from the supporting calculations.

An "invisible element" is a cycle which has the same "initial" as its "tag". (For these terms see Section 3.) Such elements are suppressed from the Curtis table because they do not contribute to the Adams spectral sequence (stable or unstable). Generically speaking, most cycles are of this kind, so their omission from the table is a great saving of space. Their omission from the calculations gives a correspondingly great saving in time.

It is important to know that these elements can be recovered whenever the information they contain is required in subsequent calculations; see Section 5.

2.7. Suppress all terms of each cycle except the leading (maximal) term, not only in the table but in the calculations. Similarly for each tag.

This is an important idea with lots of consequences. In calculations, if we are satisfied to know only the leading term of a cycle or tag, we can take many short cuts. On the other hand, when later calculations require reference to a certain cycle or tag, we may have to re-calculate its complementary terms, and this may occur many times over. In its simplest terms this is a trade-off between time and storage. If we carry only leading terms, we save storage, but lose time in repetition of certain steps (completing certain cycles). However, we also gain back some time from the short cuts used where the leading term is quickly found and suffices.

Hansen used this idea, but later came to question its efficacy. Salomonsen carried all terms. I have written programs both ways and have not yet decided how useful this idea will ultimately prove. I have one program which finds and stores the leading term only in the normal course, but when subsequent calculations refer to the full cycle, finds and stores the full cycle correctly. This uses more storage than leading-term-only methods and it is not clear whether in the long run it will be faster.

The problem of recovering a full cycle from its leading term is interesting and important and will be discussed in Section 6.

2.8. Suppress, in the table and the calculations, the information about whether a cycle actually bounds or is only homologous to a sum of smaller cycles.

For most purposes it is enough to know that a cycle is not independent of the smaller basis cycles. As an example, the boundary of 9 is 71 + 53. One may decide to carry only 71/9 and not 71/9:53. One keeps the information that 71 is no longer independent (no longer in the minimal basis) beginning with S^{10}, and suppresses the information that 71 does not bound but becomes homologous to the nonbounding cycle 53.

Ideas 2.7 and 2.8 go together in practice. If one carries full cycles, it is easy enough to carry all components; if one carries leading terms only, it is natural to neglect all the components of a boundary other than the maximal one.

A method of generating a Curtis table which uses idea 2.1 through 2.6 I will call a "Curtis method." A method using 2.1 through 2.5 but carrying invisible elements internally may be called a "full-basis method" since internally one carries the full basis for cycles at each bi-grading. A full-basis method is sketched in Section 4. A Curtis method using 2.7 will be called a "leading-term-only method" as opposed to a "full-cycle method."

3. Example and terminology

Consider the bi-grading $r = 8$, $s = 3$ $(t = 11)$. Here Λ has vector-space rank 17, but Λ' has rank only six; the other elements end with an even index. A detailed Curtis table might contain the following information:

$$A = 611 + 341 \;/\; 81 + 45 \;:\; 233$$

$$B = 521 + 323 \;/\; 63$$

$$C = 431 + 341 \;/\; 45 \;:\; 233$$

$$D = 233$$

The ordered minimal cycle basis is given by A, B, C, and D (the letters are merely for reference in this discussion). Stably, the relative basis for homology is given by D, but unstably A and B also appear; e.g., B is in the relative basis for $\Lambda(6)$, i.e., for S^6. The display indicates that 63 is the smallest chain that has B as the maximal component of its boundary. We say that 63 is the "tag" of B. The table also shows that 45 is the tag of C and in this case it displays the other component of $d(45)$, namely D.

By the "initial" of a polynomial we mean the first factor of its maximal term. Thus C and its tag both have initial 4, which means that C is an "invisible element".

When a cycle is written as a sum of basis cycles we call those the "components" of the cycle.

In the published Curtis tables, invisible elements, non-maximal components, and all terms other than leading terms are all suppressed. Thus the display is abbreviated to

611/81
521/63
233

This condensation retains explicitly all the information about the additive structure of the E_2 term of the stable and unstable Adams spectral sequence, and about Hopf invariant and sphere of origin. Furthermore, all the information in the original display is recoverable from the condensed table, by methods indicated in Section 6.

In the short listing we have only 3 of the original 17 monomials. In a sense these are the only important ones for our purpose. The idea of Curtis methods is to see how far we can go toward suppressing the other 14 monomials completely. Of course in higher dimensions the condensation becomes more and more dramatic.

. <u>The minimal basis algorithm.</u>

In this section we sketch an algorithm for finding the homology of a graded differential algebra using the idea 2.3 of ordering everything and finding a minimal basis for cycles modulo boundaries. This will be stated in terms of the mod 2 lambda algebra but is easily seen to be applicable in a much more general setting. The process also finds the tag for each cycle in the minimal basis (if any). Thus when applied to Λ the process generates a Curtis table, obtained by deleting invisible elements from the cycle bases.

In a minimal cycle basis every cycle has a different leading term. Thus each basis cycle is determined by, and can be referred to by, its leading term. The group of cycles may be regarded as the row space of a matrix, where the columns correspond to the ordered monomial basis for the chain group; then the minimal basis corresponds to the rows in the row-echelon form.

By a basis for cycles modulo boundaries (or a "relative" basis) we mean a set of cycles which taken together with the boundaries spans the cycle group, and which is minimal for this property.

By a "box" we mean a bi-grading. For the algorithm, order the boxes lexicographically in t and r. Calculation of a box is done in two steps ("d out" and "d in"). After the first step we have a cycle basis; after the second, a relative basis.

Suppose we have done the first step for the box at (r,s) and have done both steps for all prior boxes. We must now study the boundary d from $(r+1,s-1)$ to (r,s). Call these boxes "source" and "target". List the monomial basis for the chains at source, in order. We will process these monomials one by one beginning with the smallest.

Let x be the next largest monomial.

If $d(x) = 0$, adjoin x to the cycle basis at source. Otherwise write $d(x)$ as sum of basis cycles at target. Let z denote the largest of these components. If z has not already been tagged, then x tags z. If z is already tagged by w, replace x by $x' = x + w$, replace $d(x)$ by $d(x')$, and re-enter this paragraph with x' in lieu of x.

The recursive process described in the preceding paragraph always terminates in a finite number of steps, because the leading terms of $d(x)$ and $d(w)$ are the same, namely the leading term of z, so the largest component of $d(x)$ is driven downwards. The result of the process is either that x is completed to a cycle x^* which is adjoined to the cycle basis at source, or else that x is completed to the tag x^* of some basis cycle at target.

Continuing upward through the monomial basis at source in this manner, we eventually complete each monomial at source to a cycle or to a tag. It is not hard to prove that the process leads to the correct homology at target and the minimal cycle basis at source.

To obtain a Curtis table we simply condense the resulting lists by retaining only the basis cycles that are untagged or have tags with strictly larger initials. It is a matter of preference whether or not to include in the table full cycles and full tags and not just leading terms, and whether or not to include the non-maximal components of dependent cycles.

Writing out the monomials at (r,s) is tedious. It is best done by copying out λ_n times those basis monomials at $(r-n,s-1)$ that give admissible products with λ_n (i.e., have initial $\leq 2n$) for $n = 1,2,3,\ldots$. Moreover, considerable repetition in calculation is avoided if one copies complete cycles and tags and not just leading terms.

The algorithm sketched above is not particularly efficient. Its virtue is that it is easy to describe and easy to prove. Moreover it illustrates the kind of recursion which is essential to the more complicated and more efficient Curtis methods.

5. Suppressing invisible elements

It is easy to prove from the results in Wang's paper [7] that if x is tagged by y, and if they have the same initial n, then $(\lambda_i x)^*$ is tagged by $(\lambda_i y)^*$ for all $i \geq n/2$, where the asterisk indicates completion as obtained by the process in Section 4.

We can modify the algorithm of Section 4 by copying only the leading terms in such cases, and not completing unless full cycle or full tag are called for later in the calculations. Better yet, we can omit these elements entirely. If a monomial is missing at a later stage, it will be because of this deliberate omission, which can easily be traced back. For example, if one does not find the cycle beginning 631, one "divides by λ_6" and finds 31/5, indicating that 631 is tagged by 65 (in terms of leading terms only). In this example 631 itself is not a complete cycle; if the other terms are needed they may be obtained by going back to the full process as described in Section 4 for 631 and for any other elements left uncompleted and required for the full routine for 631.

6. Leading terms only

It is often possible to anticipate the outcome of the recursive process before it has terminated. One way this occurs is in connection with the invisible elements, as in Section 5. Another way depends on another structural property of Λ, namely the easily proved fact that $d(x)$ is always lexicographically less than x. If we are studying x in a case where all the target basis cycles that are lexicographically less than x (or less than $d(x)$) have already been tagged, then $d(x)$ is necessarily dependent on those cycles and thus x must complete to a cycle. We are thus led to enter x as a cycle at source -- without completing it -- and to go on to the next listing. This is one of the many ways we are led to use the leading term as the "name" of the cycle.

Our abbreviation to leading term only interferes with the logic of the recursive process. However, Curtis observed that a modification of the recursive process appears to work if we systematically use only the leading terms of cycles and tags.

On the surface the description of the algorithm and of its key recursive process is unchanged from the presentation in Section 4, but there is a subtle difference which affects the proof. In Section 4 we used full tags; now we use only the leading term of the tag. This means that the differential of the (monomial) tag may fail to contain the leading term of the tagged cycle. For example, the tag of 433 at (10,3) is 10.1 + 83 + 65; d(10.1) does not contain 433. Thus we can no longer assert that $d(x')$ is smaller than $d(x)$, and the finiteness proof fails. We will encounter cases in the calculations where the leading term of $d(x)$ becomes larger, recurs, recurs with larger complement, etc. (Such phenomena appear in working out 5.10.1.1.1 or 18.1.1.1.)

To prove finiteness of the algorithm in this setting, instead of the usual method of proof, which is to show that some discrete positive quantity is driven toward zero (as in the proof of the classical Euclidean algorithm), we must show that each step of the recursion deletes one or more elements from a finite set which thus is eventually exhausted. The set in question is the set of all terms of all complete tags of all target cycles smaller than the largest component of $d(x)$. To prove finiteness one must show that each monomial tag which arises in the processing of x belongs to that set, and will not appear in the processing more often than it appears in that set. This is not hard when the idea is at hand. The proof of correctness is rather straightforward.

The above recursion may be stopped when the outcome (cycle or tag) is clear, or it may be continued to the end, in which case one obtains the full cycle or full tag and has the other components available. These components are just those smaller target basis cycles which remain in $d(x)$ when the recursion stops because the largest component is not already tagged. This information may be kept or discarded according to the ultimate purpose of the calculation.

7. Current status

We have indicated several ideas for organizing the calculation of a Curtis table. The two methods described lie at opposite extremes: that of Section 4 is a full-basis, full-cycle method; that of Section 6 is a leading-term-only Curtis method. The first maximizes retention and storage of information, while the second represents maximum abbreviation and minimum storage. The trade-off is difficult to evaluate in advance; as the calculation moves into higher dimensions it grows very rapidly, and differential growth rates are critical for long-term efficiency. For this reason I have been experimenting at the machine with several different methods. The most promising was mentioned in Section 4; it sets out to abbreviate as much as possible, ut in cases where a complicated situation must be worked through, it retains the etails so that in an eventual recurrence this part of the calculation need not be epeated.

Programs are run on the IBM 370 at the UICC Computer Center. To date they ave been written in Snobol, more for convenience than for efficiency. Three al- orithms have been written (essentially those already mentioned) and two have been roved correct and finite. All three have been programmed and debugged in various ersions for the prime 2, and the main effort now is to improve efficiency. For the rimes 3 and 5 some versions are partially programmed.

Present versions are not efficient enough at the prime 2 to give new re- ults without using many hours of machine time, but the use of the machine for uxiliary calculations has made possible some results which it seemed unreasonable o attempt by hand; some examples have been presented elsewhere [6]. I believe the ıethod, with some theoretical modifications, will do more for us at the prime 3.

REFERENCES

[1] Bousfield, Curtis, Kan, Quillen, Rector, and Schlesinger, Topology 5 (1966) 331-342. MR 33 #8002.

[2] Bousfield, A.K., and D.M. Kan, The homotopy spectral sequence etc., Topology 11 (1972) 79-106, especially pp. 101-102. MR 44 #1031.

[3] Curtis, E.B. Simplical homotopy theory. Lecture notes, Aarhus Universitet, 1967. MR 42 #3785. Reprinted, slightly revised and enlarged, in Advances in Math. 6 (1971) 107-209. MR 43 #5529. Curtis table on p. 104 of Aarhus notes (to the 23-stem) and p. 190 of Advances (to the 16-stem).

[4] Hansen, Wm. A., Computer calculation of the homology of the lambda algebra. Dissertation, Northwestern University, 1974.

[5] Tangora, M.C., On the cohomology of the Steenrod algebra. Dissertation, Northwestern University, 1966. Slightly revised and condensed, Math. Z. 116 (1970) 18-64. MR 42 #1112. (A presentation of some of the ideas in Section 2 was condensed out of the published version.)

[6] Tangora, M.C., Some remarks on the lambda algebra. Submitted to Proceedings of the March 1977 Conference on Topology at Evanston, Illinois.

[7] Wang, J.S.P., On the cohomology of the mod-2 Steenrod algebra etc., Ill. J. Math. 11 (1967), 480-490. MR 35 #4917.

[8] Whitehead, G.W. Recent advances in homotopy theory. Regional Conference Series (A.M.S.-Conference Board), 1970. MR 46 #8208. The table is on pp.71-73 (to the 22-stem).

FLAG MANIFOLDS AND HOMOTOPY RIGIDITY OF LINEAR ACTIONS

BY

ARUNAS LIULEVICIUS†

Let $U = U(n+1)$ be a unitary group, $T = T^{n+1}$ the maximal torus in U consisting of diagonal matrices. Our aim in this paper is to prove that if G is a compact topological group then the linear actions of G on the flag manifold U/T are rigid under homotopy. Our main tool is the following result on the structure of the group of automorphisms of a graded algebra.

Let $Z[x_1,\ldots,x_n]$ be the polynomial algebra on indeterminates x_i of grading 2. Define the element h_2 to be the sum of all distinct quadratic monomials in the x_i's. For example if $n = 2$ we have $h_2 = x_1^2 + x_1x_2 + x_2^2$. Let I be the ideal in $Z[x_1,\ldots,x_n]$ generated by h_2.

Theorem 1. If $n \geq 2$ then the group of algebra automorphisms of the graded algebra $Z[x_1,\ldots,x_n]/I$ is $S_{n+1} \times Z/2Z$, where S_{n+1} is the symmetric group on n+1 letters. Define x_{n+1} by setting $x_{n+1} = -x_1-x_2-\ldots-x_n$, then S_{n+1} acts on $x_1,x_2,\ldots,x_n,x_{n+1}$ by permuting their indices. The generator T of $Z/2Z$ acts by $Tx_i = -x_i$ for all i.

The cohomology of U/T is a quotient of the algebra $Z[x_1,\ldots,x_n]/I$. Indeed, it is classical [2] that $H^*(U/T;Z) = Z[x_1,\ldots,x_n]/J$, where $J = (h_2,\ldots,h_{n+1})$, h_m being the sum of all distinct monomials in the generators x_i in grade $2m$. The group of U-maps of U/T is the Weyl group $N_U(T)/T = S_{n+1}$. Let $c: U \to U$ be complex conjugation of matrices, $c(u) = \bar{u}$. Since $c(T) = T$ we obtain a map $\underline{c}: U/T \to U/T$.

Corollary 2. The homomorphism

$$\phi: N_U(T)/T \times Z/2Z \to \text{Aut } H^*(U/T;Z)$$

defined by $\phi(k,t) = k^*c^{t*}$ is onto the group of all algebra automorphisms of $H^*(U/T;Z)$. If $n \geq 2$ ϕ is an isomorphism.

† Research partially supported by NSF grant MCS 75-08280

Let $H \subset U$ be a closed subgroup. An action of a compact group G on U/H is said to be *linear* if there is a representation $\alpha: G \to U$ such that the action is given by $g.uH = \alpha(g)uH$. We shall denote this linear action by $(U/H,\alpha)$. We shall say that linear actions of G on U/H have the property of *homotopy rigidity* if given representations $\alpha,\beta: G \to U$ and a G-map $f: (U/H,\alpha) \to (U/H,\beta)$ with $f: U/H \to U/H$ a homotopy equivalence there exists a linear character $\chi: G \to S^1$ such that either β or $\bar{\beta}$ is similar to χ_α.

Theorem 3. Linear actions of a compact group G on U/T have the property of homotopy rigidity.

Homotopy rigidity of linear actions on $CP^n = U(n+1)/U(1)xU(n)$ has been shown in [5] and [6]. A proof that linear actions on $U(n+k)/U(n) \times T^k$ (where $n \geq k$) have the homotopy rigidity property appears in [7]. Homotopy rigidity of linear actions on $U(m+n+1)/U(m) \times U(n) \times U(1)$ is proved in [8]. A conjecture has been made in [7] that linear actions on U/H are homotopy rigid if H is a friendly subgroup of U: H is closed, connected, of maximal rank in $U = U(n+1)$ and there exists a non-zero vector v in C^{n+1} such that $hv = \lambda(h)v$ for some linear character $\lambda: H \to S^1$. The case $H = T$ is central - one hopes that the proof of the general conjecture is close at hand.

The paper is organized as follows: section 1 presents a proof of Theorem 1; section 2 shows how Theorem 1 implies Corollary 2 and Theorem 3.

The author wishes to thank J. Alperin for a key idea and an exhortation to think geometrically. Thanks also go to G. Glauberman, I. Kaplansky, R. Narasimhan, and R. Stong for their helpful comments.

1. Proof of Theorem 1: Let V be a vector space over the real numbers with basis $\{e_1,\ldots,e_n\}$. We define a bilinear pairing $H: V \times V \to R$ by setting $H(e_i,e_i) = 1$, $H(e_i,e_j) = 1/2$ for $i \neq j$, $i,j = 1,\ldots,n$.

Lemma 4. H is an inner product on V.

The point to prove is that H is positive definite. If $(y_1,\ldots,y_n)$ are the coordinates of a vector y in V with respect to the basis $\{e_1,\ldots,e_n\}$ then

$H(y,y) = h_2(y_1,\ldots,y_n)$. We prove that H is positive definite by exhibiting $h_2(y_1,\ldots,y_n)$ as a sum of squares with positive rational coefficients.

Lemma 5. Let $a_j = (j+1)/2j$, then

$$h_2(y_1,\ldots,y_n) = \sum_{j=1}^{n} a_j \left(y_j+\frac{1}{j+1}\, y_{j+1}+\ldots+\frac{1}{j+1}\, y_n\right)^2.$$

The reader is invited to prove Lemma 5 himself. A very elegant way of proceeding is sketched in [4], exercises 6 (p.5) and 4 (p.14).

Let Γ be the integral lattice in V generated by the basis $\{e_1,\ldots,e_n\}$. We determine the elements of minimal length in Γ.

Proposition 6. Let $y \in \Gamma$ and $H(y,y) = 1$, then y is e_i, $-e_i$, or e_i-e_j for some $i \neq j$, $i,j = 1,\ldots,n$.

Proof: Notice that $H(e_1-e_2,e_1-e_2) = 1-1/2-1/2+1 = 1$, so the elements listed indeed have length 1. Now suppose that $y = y_1e_1+\ldots+y_ne_n$ with y_i integers and $H(y,y) = 1$. Let us prove that y_n is at most 1 in absolute value. We have $1 = H(y,y) = S+a_ny_n^2$, where S is positive, $a_n = (n+1)/2n$, so $a_ny_n^2 = (1+\frac{1}{n})\frac{1}{2}\, y_n^2$, hence $y_n^2 = 0$ or 1. If there is exactly one non-zero y_i then $y = e_i$ or $y = -e_i$. Suppose there are at least two non-zero y_i. We can assume (by relabeling the variables) that y_{n-1} and y_n are non-zero. We claim: they have opposite sign. Suppose not. We may as well assume that $y_{n-1} = y_n = 1$. We have

$$1 = H(y,y) = P + \frac{n}{2n-2}\left(1 + \frac{1}{n}\right)^2 + \frac{n+1}{2n},$$

where P is a positive number. Now this simplifies to $1-P = \frac{n+1}{n-1}$, so this is a serious embarassment. This contradiction shows that y_{n-1} and y_n must have opposite signs. This means in particular that there are at most two non-zero entries among the y_i and so y is e_i, $-e_i$, or e_i-e_j for $i \neq j$, as was to be shown.

Lemma 7. $H(e_1,e_1-e_2) = 1/2$, $H(e_1,e_2-e_3) = 0$, $H(e_1-e_2,e_1-e_3) = 1/2$, $H(e_1-e_2,e_3-e_4) = 0$.

Of course the lemma is to be interpreted as follows: if $n = 2$, ignore the ast three equations, if $n = 3$, ignore the last equation. If $n \geq 4$ we use S_n) replace $1,2,3,4$ by pairwise distinct indices i,j,s,t.

roposition 8. Let $T\colon V \to V$ be an R-linear transformation such that $T(\Gamma) \subset \Gamma$ and $(Tu,Tv) = H(u,v)$ for all u, v in V. Then there exists an ϵ such that $\epsilon = 1$ r -1 and one of the two following cases holds: either there exists a σ in S_n uch that $Te_j = \epsilon e_{\sigma(j)}$ for all j, or there exists an i such that $Te_j = \epsilon e_i$ or $(e_i - e_k)$ for some $k \neq i$.

emark. The point of the proposition is this: if we agree to write coordinates with espect to the basis $e_1,\ldots,e_n$ in rows then the matrix of T is either ϵ times permutation of the identity matrix or has the form

$$\begin{bmatrix} & & \epsilon & & -\epsilon \\ & & \cdot & & \\ & & \cdot & & \\ & & \cdot & & \\ -\epsilon & & \epsilon & & \end{bmatrix} \quad \text{(column } i\text{)}.$$

roof: If $Te_j = \epsilon_j e_{\sigma(j)}$ for some σ in S_n, we have to show that

$$\epsilon_1 = \epsilon_2 = \ldots = \epsilon_n ,$$

ut this is easy, for

$$1/2 = H(e_j,e_k) = H(Te_j,Te_k) = \epsilon_j\epsilon_k H(e_{\sigma(j)},e_{\sigma(k)}) = 1/2\ \epsilon_j\epsilon_k$$

for $j \neq k$, so $\epsilon_j = \epsilon_k$ as claimed. Now suppose for some j we have $Te_j = e_i - e_k$. We claim: for $s \neq j$ we either have $Te_s = e_i$, or $Te_s = -e_k$, or $Te_s = e_i - e_t$, or $Te_s = -e_k + e_t$ -- this is a consequence of Lemma 7 since $H(e_j,e_s) = 1/2$ for $j \neq s$. If $n \geq 3$ another application of Lemma 7 shows that e_i or $-e_k$ always occurs in Te_s (and this is independent of s). This completes the proof of the proposition.

Proof of Theorem 1: Let V be a vector space over the real numbers with basis $\{e_1,\ldots,e_n\}$ and we consider V^* with the dual basis $\{x_1,\ldots,x_n\}$. Let $T^*\colon V^* \to V^*$ be a linear transformation which preserves the integer lattice generated by

$\{x_1,\ldots,x_n\}$ and satisfies $T^*h_2 = a\, h_2$ with $a = 1$ or -1. Here we have used the symbol for the transformation on $R[x_1,\ldots,x_n]$ induced by T^*. It follows that the dual of T^* (which we of course denote by T) preserves the integral lattice Γ in V and satisfies $H(Tu,Tv) = aH(u,v)$ for all u, v in V. Thus $a = 1$ and therefore T is a transformation to which we can apply Proposition 8. If we write the coordinates in V^* with respect to the basis $\{x_1,\ldots,x_n\}$ in columns, then the matrix of T^* with respect to this basis is the same as the matrix of T with respect to this basis $\{e_1,\ldots,e_n\}$ (if we write the coordinates with respect to this basis in rows). Hence if we let $x_{n+1} = -x_1-\ldots-x_n$ the remark after Proposition 8 gives T^* as an element of $S_{n+1} \times Z/2Z$. This proves Theorem 1.

2. Homotopy rigidity for U/T.

We first show how Theorem 1 implies Corollary 2.

Consider the standard fibration [2]

$$\begin{array}{ccc} U/T & \longrightarrow & BT \\ & & \downarrow \\ & & BU. \end{array}$$

The inclusion of the fiber induces an epimorphism on integral cohomology. Let λ_i be the line bundle on BT induced by the i-th coordinate projection $\pi_i\colon T \to S^1$ and $x_i = c_1(\lambda_i)$ be its first Chern class. Then $H^*(BT;Z) = Z[x_1,\ldots,x_{n+1}]$ and

$$\begin{aligned} H^*(U/T;Z) &= Z[x_1,\ldots,x_{n+1}]/(h_1,h_2,\ldots,h_{n+1}) \\ &= Z[x_1,\ldots,x_n]/(h_2,\ldots,h_{n+1}), \end{aligned}$$

where the reader should not be confused by our two uses of h_m -- in the first case we have $h_m = h_m(x_1,\ldots,x_{n+1})$, and in the second $h_m = h_m(x_1,\ldots,x_n)$. Under the isomorphism, x_{n+1} of course corresponds to $-x_1-\ldots-x_n$. Since the quotient map

$$Z[x_1,\ldots,x_n]/I \to H^*(U/T;Z)$$

is an isomorphism for grades ≤ 5 (remember: I is the principal ideal generated by h_2) this means that the group of algebra automorphisms of $H^*(U/T;Z)$ is a subgroup of the group of algebra automorphisms of $Z[x_1,\ldots,x_n]/I$. We claim: this inclusion is the identity map - that is we have to show that each of the elements of $S_{n+1} \times Z/2Z$ comes from an automorphism of $H^*(U/T;Z)$. We shall show even more: each element of this group is induced by a map of U/T into U/T of a certain kind. The group of U-maps of U/T into U/T is the Weyl group $N_U(T)/T = S_{n+1}$ and it acts on $H^*(BT;Z) = Z[x_1,\dots,x_{n+1}]$ by permutation of the x_i. Conjugation $c: U \to U$ induces $c: T \to T$ and $Bc: BT \to BT$ which in cohomology becomes $Bc^*x_i = -x_i$ for $i = 1,\dots,n+1$. Thus each element of $S_{n+1} \times Z/2Z$ corresponds to an automorphism of $H^*(U/T;Z)$ induced by a U-map $U/T \to U/T$ possibly followed by conjugation $c: U/T \to U/T$. This shows that if $n \geq 2$ the map ϕ is an isomorphism, and if $n = 1$ ϕ is onto with kernel being $Z/2Z$ (the diagonal of $S_2 \times Z/2Z$). This proves Corollary 2.

The reader should compare this argument with [3] -- there cohomology endomorphisms of Grassmann manifolds are also determined by the bottom relation.

Our task is now to show how Corollary 2 implies Theorem 3. Let $\gamma: G \to U$ be a representation. The Hopf bundle $h(\gamma): (S^{2n+1},\gamma) \to (CP^n,\gamma)$ determines an element $h(\gamma)$ in $Pic_G(CP^n,\gamma)$, the Picard group of G-equivariant complex line bundles over (CP^n,γ). The inclusion $T \subset U(1) \times U(n)$ gives rise to a G-equivariant map $\pi_\gamma: (U/T,\gamma) \to (CP^n,\gamma)$. Suppose we are given two representations $\alpha,\beta: G \to U$ and a G-map $f: (U/T,\alpha) \to (U/T,\beta)$ such that $f: U/T \to U/T$ is a homotopy equivalence. According to Corollary 2 we can find an element k in $N_U(T)/T \times Z/2Z$ such that k^*f^* is the identity map on $H^*(U/T;Z)$. We replace f by fk (and possibly α by $\bar{\alpha}$ if $\underline{c}$ is involved in k) to obtain a G-map $f: (U/T,\alpha) \to (U/T,\beta)$ with f^* = identity map of $H^*(U/T;Z)$. Now consider the equivariant K-theory functor K_G (see [1], [9]), then

$$\pi_\gamma^!: K_G(CP^n,\gamma) \to K_G(U/T,\gamma)$$

is a monomorphism (see [9]) and $K_G(CP^n,\gamma)$ is a free $K_G(\text{point}) = R(G)$ module on $\{1,h,\dots,h^n\}$, where $h = h(\gamma)$ and

$$h^{n+1}-\gamma h^n+\Lambda^2\gamma h^{n-1}-\dots+(-1)^{n+1}\Lambda^{n+1}\gamma = 0 .$$

In our situation above let $s = h(\alpha)$, $t = h(\beta)$, then the condition f^* = identity gives us $i^!f^!\pi_\beta^!t = i^!\pi_\alpha^!s$, where $i: E \to G$ is the inclusion of the identity subgroup E into G. Since U/T is simply connected every G-equivariant line bundle over $(U/T,\alpha)$ which is trivial as an ordinary line bundle over U/T is induced from a G-line bundle over a point via a collapsing map $J: U/T \to *$ (see Theorem 2 of [7] -- the proof uses G. Segal's technique of cohomology of topological groups -- see [10]). This means that there is a linear character $\chi: G \to S^1$ such that $f^!\pi_\beta^! t = \pi_\alpha^!(\chi s)$. Since $\pi_\alpha^!$ is a monomorphism we define a map of R(G)-algebras

$$\psi: K_G(CP^n,\beta) \to K_G(CP^n,\alpha)$$

by setting $\psi = (\pi_\alpha^!)^{-1} f^! \pi_\beta^!$, so $\psi(t) = \chi s$. We use the expansion of t^{n+1} in terms of $1, t, \ldots, t^n$, apply ψ to it, multiply by χ^{-n-1}, and compare this expansion of s^{n+1} with the standard expansion. Looking at the coefficient of s^n in the two expansions we obtain $\beta = \chi\alpha$ which is precisely Theorem 3.

If the reader wishes to see a more leisurely discussion of this method of proof, please see [7].

REFERENCES

[1] M.F. Atiyah and G.B. Segal, Lectures on equivariant K-theory, Mimeographed notes, Oxford 1965.

[2] A. Borel, Sur la cohomologie des espaces fibrés principaux et des espaces homogènes de groupes de Lie compacts, Annals of Math. 57 (1953), 115-207.

[3] H. Glover and W. Homer, Endomorphisms of the cohomology rings of finite Grassmann manifolds, Proceedings of the Northwestern University homotopy theory conference, March 1977 (to appear).

[4] I. Kaplansky, Linear Algebra and Geometry, A Second Course, 2nd edition, Chelsea Publishing Company, New York, 1974.

[5] A. Liulevicius, Homotopy types of linear G-actions on complex projective spaces. Matematisk Institut, Aarhus Universitet, Preprint Series 1975/76, No. 14.

[6] ____________, Characters do not lie. Transformation Groups (ed. Czes Kosniowski), Proceedings of the conference on Transformation Groups, Newcastle upon Tyne, August 1976, Cambridge University Press (1976), 139-146.

[7] ____________, Homotopy rigidity of linear actions: characters tell all (to appear in the Bulletin AMS).

[8] ____________, Line bundles, cohomology automorphisms, and homotopy rigidity of linear actions, Proceedings of the Northwestern University homotopy theory conference, March 1977 (to appear).

[9] G.B. Segal, Equivariant K-theory, Publ. Math. I.H.E.S. 34 (1968), 129-151.

10] __________, Cohomology of topological groups, Symposia Mathematica, vol. IV (INDAM, Rome, 1968/69), 377-387.

GENERALIZED HOMOLOGICAL REDUCTION THEOREMS*

By

Denis Sjerve

§1 Introduction:

In this paper we shall give topological proofs of some reduction theorems in homological algebra(see p.228 of [1]). Thus suppose given a group π and an exact sequence of left π modules

(E) $$0 \longrightarrow A \xrightarrow{\mu} P_{n-1} \xrightarrow{\partial_{n-1}} P_{n-2} \xrightarrow{\partial_{n-2}} \cdots \xrightarrow{\partial_1} P_0 \xrightarrow{\varepsilon} Z \longrightarrow 0$$

where the P_i are projective and Z has the trivial module structure. For the exact sequence (E) we have the iterated connecting homomorphism $\delta:\mathrm{Hom}_\pi(A,A) \longrightarrow \mathrm{Ext}^n_\pi(Z,A)$. Thus there exists a distinguished cohomology class $\Delta \varepsilon H^n(\pi;A) \cong \mathrm{Ext}_\pi(Z,A)$ corresponding to the identity $1_A:A \longrightarrow A$. If B,C are left π modules and if $B \otimes A$, Hom(A,C) are made into left π modules via the usual diagonal actions then the reduction theorems say that the homomorphisms

(1.1) $$H_{s+n}(\pi;B) \longrightarrow H_s(\pi;B \otimes A),\ u \longrightarrow u \cap \Delta$$

(1.2) $$H^s(\pi;\mathrm{Hom}(A,C)) \longrightarrow H^{s+n}(\pi;C)\ ,\ u \longrightarrow u \cup \Delta$$

are isomorphisms for s>0. In (1.2) we are using the evaluation pairing $\mathrm{Hom}(A,C) \otimes A \longrightarrow C$. These homomorphisms are not isomorphisms for s=0. But, by identifying $H_0(\pi;B \otimes A)$ with $B \otimes_\pi A$ and $H^0(\pi;\mathrm{Hom}(A,C))$ with $\mathrm{Hom}_\pi(A,C)$ we have instead the exact sequences

(1.3) $$0 \longrightarrow H_n(\pi;B) \xrightarrow{\cdot \cap \Delta} B \otimes_\pi A \xrightarrow{1 \otimes \mu} B \otimes_\pi P_{n-1}$$

(1.4) $$\mathrm{Hom}_\pi(P_{n-1},C) \xrightarrow{\mathrm{Hom}(\mu,1)} \mathrm{Hom}_\pi(A,C) \xrightarrow{\cdot \cup \Delta} H^n(\pi;C) \longrightarrow 0$$

The first thing to be noticed is that the groups $H_{s+n}(\pi;B)$, $H^{s+n}(\pi;C)$ do not depend on the exact sequence (E) ; whereas, the groups $H_s(\pi;B \otimes A)$, $H^s(\pi;\mathrm{Hom}(A,C))$ and the class $\Delta \varepsilon H^n(\pi;A)$ certainly seem to. However, if we are given another such

---

* Research partially supported by N.R.C. Contract A 7218.

xact sequence

E') $$0 \longrightarrow A' \longrightarrow P'_{n-1} \longrightarrow \cdots \longrightarrow P'_o \longrightarrow Z \longrightarrow 0$$

hen a Schanuel type lemma (see [7])implies that there are projectives Q,Q' such

hat $A \oplus Q \cong A' \oplus Q'$. Now $H_s(\pi;B \otimes Q) \cong 0$ for $s>0$ since $B \otimes Q$ is a relative projective, and

ence we have for $s>0$

$$H_s(\pi;B\otimes(A\oplus Q)) \cong H_s(\pi;B\otimes A) \oplus H_s(\pi;B\otimes Q) \cong H_s(\pi;B\otimes A)$$

ikewise $H_s(\pi;B\otimes(A'\oplus Q')) \cong H_s(\pi;B\otimes A')$ for $s>0$. Since $A\oplus Q \cong A'\oplus Q'$ it follows that

or $s>0$ $H_s(\pi;B\otimes A) \cong H_s(\pi;B\otimes A')$, and so these groups do not depend on (E). Moreover

f $\Delta' \varepsilon H^n(\pi;A')$ is the cohomology class corresponding to the identity $1_{A'}:A' \longrightarrow A'$

ia the connecting homomorphism $\delta:\mathrm{Hom}_\pi(A',A') \longrightarrow \mathrm{Ext}^n_\pi(Z,A') \cong H^n(\pi;A')$, then for

>0 we have the commutative diagram

$$H_{s+n}(\pi;B) \xrightarrow{\cdot\cap\Delta} H_s(\pi;B\otimes A)$$
$$H_{s+n}(\pi;B) \xrightarrow{\cdot\cap\Delta'} H_s(\pi;B\otimes A')$$
$$H_s(\pi;B\otimes A) \cong H_s(\pi;B\otimes A')$$

n other words, the homological reduction isomorphism (1.1) does not depend on (E).

ikewise (1.2) does not depend on the exact sequence (E).

The second thing to be noticed is that there is no loss of generality if we

ssume all the P_i are free π modules. To see this first pick a projective Q_0 such

that $P_0 \oplus Q_0 = F_0$ is free. Then replace (E) by

(E_0) $$0 \longrightarrow A \longrightarrow P_{n-1} \longrightarrow \cdots \longrightarrow P_2 \longrightarrow P_1 \oplus Q_0 \longrightarrow P_0 \oplus Q_0 \longrightarrow Z \longrightarrow 0$$

Now choose a projective Q_1 such that $P_1 \oplus Q_0 \oplus Q_1 = F_1$ is free and replace (E_0) by

(E_1) $$0 \longrightarrow A \longrightarrow P_{n-1} \longrightarrow \cdots \longrightarrow P_3 \longrightarrow P_2 \oplus Q_1 \longrightarrow P_1 \oplus Q_0 \oplus Q_1 \longrightarrow P_0 \oplus Q_0 \longrightarrow Z \longrightarrow 0$$

Iterating this procedure we arrive at an exact sequence

(E_{n-1}) $$0 \longrightarrow A \oplus Q \longrightarrow F_{n-1} \longrightarrow \cdots \longrightarrow F_0 \longrightarrow Z \longrightarrow 0$$

where the F_i are free and Q is projective. But, as we observed above, the

reduction isomorphisms for (E_{n-1}) are equivalent to those for (E).

The upshot of the last two paragraphs is that to prove the reduction theorems

we need only consider a particular exact sequence. For example, we may take the

truncated bar resolution

(1.5) $$0 \longrightarrow A \longrightarrow \overline{B}_{n-1}(\pi) \longrightarrow \cdots \longrightarrow \overline{B}_0(\pi) \longrightarrow Z \longrightarrow 0$$

But topology now enters into the picture since this resolution has a geometric description. Specifically, if W is the iterated join $\pi*\pi*\ldots*\pi$(n copies) then W is an n-1 dimensional, n-2 connected simplicial complex with a simplicial free properly discontinuous left action by π. The chain groups $C_i(W)$, $0\leq i\leq n-1$, are then free left π modules and identical to the $\bar{B}_i(\pi)$. Thus the exact sequence (1.5) becomes

$$0\longrightarrow A\longrightarrow C_{n-1}(W)\longrightarrow\cdots\longrightarrow C_0(W)\longrightarrow Z\longrightarrow 0$$

and A is the left π module $\tilde{H}_{n-1}(W)$.

The action of π on W extends in a natural way to the pair (CW,W), where CW is the cone on W. Thus the chain complex $C_*(CW,W)$ becomes a chain complex of left π modules. If B is a left π module then the chain complex $B\otimes C_*(CW,W)$ consists of π modules and therefore $H_*(CW,W;B)$ consists of π modules. In particular A is $H_n(CW,W)$ and, by the universal coefficient theorem, we have an isomorphism of π modules $B\otimes A\cong H_n(CW,W;B)$. Thus the homological reduction isomorphism becomes

(1.6) $$H_{s+n}(\pi;B)\cong H_s(\pi;H_n(CW,W;B)) \text{ for } s>0$$

But now the group $H_s(\pi;H_n(CW,W;B))$ is looking suspiciously like the E^2 term of a spectral sequence and the isomorphism (1.6) is suggesting that this spectral sequence collapses and converges to $H_{s+n}(\pi;B)$, at least for $s>0$. A similar situation exists in the cohomological case.

In fact such a spectral sequence exists in a wider generality-see [5], [6] for some of the details. The generalized reduction theorems are proved in §2, and in §3 we give some applications.

§2 The Reduction Isomorphisms:

Throughout this section we shall assume that W is an arbitrary CW complex and that π is a group with a cellular left action on W. Moreover, we shall assume that the projection $W\longrightarrow W/\pi=X_1$ is a principal π bundle. Then the preamble of the introduction suggests that the correct formulation, and generalization, of the isomorphism (1.1) is to be found in a spectral sequence whose E^2 term is

$$E^2_{s,t}\cong H_s(\pi;H_t(CW,W;B))$$

To construct this spectral sequence we extend the action of π to all the terated joins $W^k = W * \ldots * W$ (k copies) in the usual manner. This gives principal bundles $W^k \longrightarrow W^k/\pi = X_k$ for $1 \le k < \infty$. The inclusion $W^k \rightarrowtail W^{k+1}$ defined by mapping W^k onto the first k coordinates of W^{k+1} is π equivariant and so induces an inclusion $X_k \rightarrowtail X_{k+1}$. Defining W^∞, X_∞ to be the respective limits we then have a commutative diagram of inclusions and principal π bundles

$$\begin{array}{ccccccc} W & \rightarrowtail & W^2 & \rightarrowtail & \cdots & \rightarrowtail & W^\infty \\ \downarrow & & \downarrow & & & & \downarrow \\ X_1 & \rightarrowtail & X_2 & \rightarrowtail & \cdots & \rightarrowtail & X_\infty \end{array}$$

Since W^∞ is contractible the principal π bundle $W^\infty \longrightarrow X_\infty$ is universal. In particular X_∞ is a $K(\pi,1)$.

The group π acts on W^∞ and on the pair (CW,W) so we can define a fibre bundle pair

$$\omega \qquad (CW,W) \longrightarrow (D(\omega),S(\omega)) \longrightarrow X_\infty$$

by the Borel mixing construction:

the total pair is $(D(\omega),S(\omega)) = W^\infty \times_\pi (CW,W)$

the projection to X_∞ is the first coordinate function

Then one of the main theorems of [5], [6] was the existence of a functorial homeomorphism $D(\omega)/S(\omega) \cong X_\infty/X_1$; where functoriality means that this homeomorphism is natural with respect to morphisms from one principal bundle to another. Replacing $(D(\omega),S(\omega))$ by (X_∞,X_1) and then taking the Serre spectral sequence of ω will give us the spectral sequence we are seeking.

From now on all modules will be left modules and all actions will be left actions. The tensor product and hom functors will always be equipped with the diagonal actions. Since π acts on the pair (W^∞,W) the cellular chain complex $C_*(W^\infty,W)$ is a chain complex of π modules. Then we make the following definitions:

$$H_*(X_\infty,X_1;B) = H_*(B \otimes_\pi C_*(W^\infty,W)); \; H^*(X_\infty,X_1;C) = H^*(\mathrm{Hom}_\pi(C_*(W^\infty,W),C))$$

where B,C are arbitrary π modules. $H_*(X_1;B)$, $H_*(X_\infty;B)$, etc. are defined in a similar way.

There is also a π action on the pair (CW,W) and therefore the ordinary homology and cohomology groups $H_*(CW,W;B)$, $H^*(CW,W;C)$ become π modules. Then the relative Serre spectral sequences, with local coefficients, of ω are

(2.1)
$$E^2_{s,t} \cong H_s(X_\infty; H_t(CW,W;B)) \Rightarrow H_{s+t}(X_\infty, X_1; B)$$
$$E_2^{s,t} \cong H^s(X_\infty; H^t(CW,W;C)) \Rightarrow H^{s+t}(X_\infty, X_1; C)$$

Moreover, these spectral sequences are natural functors on the category of principal bundles.

In order to facilitate the study of these spectral sequences we shall assume the following hypothesis for the remainder of this section.

The Vanishing Hypothesis: W is n-1 dimensional and n-2 connected for some $n \geq 1$.

Then the only possible non-zero π module $H_t(CW,W)$ is $A=H_n(CW,W)$. Thus the universal coefficient theorems give π module isomorphisms

$$H_n(CW,W;B) \cong B \otimes A, \quad H^n(CW,W;C) \cong \mathrm{Hom}(A,C)$$

Since A is a torsion free abelian group all other π modules $H_t(CW,W;B)$, $H^t(CW,W;C)$ are zero and therefore the spectral sequences (2.1) collapse in a very strong sense, namely

(2.2) Theorem: Assuming the vanishing hypothesis we have isomorphisms for all s:

$$H_{s+n}(X_\infty, X_1; B) \cong H_s(X_\infty; B \otimes A), \quad H^{s+n}(X_\infty, X_1; C) \cong H^s(X_\infty; \mathrm{Hom}(A,C))$$

In particular $H_{s+n}(X_\infty, X_1; B)=0$ and $H^{s+n}(X_\infty, X_1; C)=0$ for $s<0$. Since X_∞ is a $K(\pi,1)$ and X_1 is n-1 dimensional (2.2) immediately gives

(2.3)
$$H_n(X_\infty, X_1; B) \cong B \otimes_\pi A, \quad H^n(X_\infty, X_1; C) \cong \mathrm{Hom}_\pi(A,C)$$

(2.4)
$$H_{s+n}(\pi; B) \cong H_s(\pi; B \otimes A), \quad H^{s+n}(\pi; C) \cong H^s(\pi; \mathrm{Hom}(A,C)) \text{ for } s>0$$

Now $A=H_n(CW,W)$ fits into the exact sequence

$$0 \longrightarrow A \xrightarrow{\mu} C_{n-1}(W) \xrightarrow{\partial n-1} \cdots \xrightarrow{\partial_1} C_0(W) \xrightarrow{\varepsilon} Z \longrightarrow 0$$

and so we see that the isomorphisms (2.4) are analogous to those of the introduction. Their product nature will be proved later in this section.

To derive exact sequences similar to those of (1.3) and (1.4) we consider portions of the exact sequences in homology and cohomology for the pair (X_∞, X_1). For homology we have

$$0 \longrightarrow H_n(X_\infty; B) \longrightarrow H_n(X_\infty, X_1; B) \xrightarrow{\partial} H_{n-1}(X_1; B) \longrightarrow H_{n-1}(X_\infty; B) \longrightarrow 0$$

But $H_*(X_1;B)$ is, by definition, computed from the chain complex

$$0 \longrightarrow B\otimes_\pi C_{n-1}(W) \longrightarrow \cdots \longrightarrow B\otimes_\pi C_0(W) \longrightarrow 0$$

and so there exists a monomorphism $H_{n-1}(X_1;B) \rightarrowtail B\otimes_\pi C_{n-1}(W)$. Replacing $H_n(X_\infty,X_1;B)$ by $B\otimes_\pi A$ gives the exact sequence

$$0 \longrightarrow H_n(\pi;B) \longrightarrow B\otimes_\pi A \xrightarrow{1\otimes\mu} B\otimes_\pi C_{n-1}(W) \tag{2.5}$$

exactly as in (1.3).

In cohomology we have the exact sequence

$$0 \longrightarrow H^{n-1}(X_\infty;C) \longrightarrow H^{n-1}(X_1;C) \xrightarrow{\delta} H^n(X_\infty,X_1;C) \longrightarrow H^n(X_\infty;C) \longrightarrow 0$$

Now we compute $H^*(X_1;C)$ from the cochain complex

$$0 \longleftarrow \mathrm{Hom}_\pi(C_{n-1}(W),C) \longleftarrow \cdots \longleftarrow \mathrm{Hom}_\pi(C_0(W),C) \longleftarrow 0$$

and consequently there exists an epimorphism $\mathrm{Hom}_\pi(C_{n-1}(W),C) \twoheadrightarrow H^{n-1}(X_1;C)$. Since $H^n(X_\infty,X_1;C) \cong \mathrm{Hom}_\pi(A,C)$ we have the exact sequence

$$\mathrm{Hom}_\pi(C_{n-1}(W),C) \xrightarrow{\mathrm{Hom}(\mu,1)} \mathrm{Hom}_\pi(A,C) \longrightarrow H^n(\pi;C) \longrightarrow 0 \tag{2.6}$$

just as in (1.4).

Finally, to complete the proofs of the generalized reduction theorems we must exhibit the product structure of the homomorphisms

$$H_n(\pi;B) \longrightarrow B\otimes_\pi A,\quad \mathrm{Hom}_\pi(A,C) \longrightarrow H^n(\pi;C)$$

in (2.5), (2.6) respectively; and of the isomorphisms in (2.4).

By (2.3) there is an isomorphism $H^n(X_\infty,X_1;A) \cong \mathrm{Hom}_\pi(A,A)$. Thus let $T \in H^n(X_\infty,X_1;A)$ be the cohomology class corresponding to the identity $1_A: A \longrightarrow A$. Then we shall prove that the isomorphisms of (2.2) are given by

$$H_{s+n}(X_\infty,X_1;B) \xrightarrow[\cong]{\cdot\cap T} H_s(X_\infty;B\otimes A) \text{ for all } s$$

$$H^s(X_\infty;\mathrm{Hom}(A,C)) \xrightarrow[\cong]{\cdot\cup T} H^{s+n}(X_\infty,X_1;C) \text{ for all } s$$

where, in the cohomological case, we are using the coefficient pairing

$$\mathrm{Hom}(A,C)\otimes A \longrightarrow C,\quad f\otimes a \longrightarrow f(a)$$

To prove this consider the commutative diagram of fibre bundles

$$\begin{array}{ccccc} (CW,W) & \longrightarrow & (D(\omega),S(\omega)) & \longrightarrow & X_\infty \\ \cup & & \cup & & \| \\ CW & \longrightarrow & D(\omega) & \longrightarrow & X_\infty \end{array} \tag{2.7}$$

Suppose M, M', M'' are π modules and $\phi: M\otimes M' \longrightarrow M''$ is some coefficient pairing. We now have three Serre spectral sequences

$$E_2^{s,t} \cong H^s(X_\infty;H^t(CW;M)) \Longrightarrow H^{s+t}(X_\infty;M)$$

$$F_2^{s',t'} \cong H^{s'}(X_\infty;H^{t'}(CW,W;M')) \Longrightarrow H^{s'+t'}(X_\infty,X_1;M')$$

$$G_2^{s'',t''} \cong H^{s''}(X_\infty;H^{t''}(CW,W;M'')) \Longrightarrow H^{s''+t''}(X_\infty,X_1;M'')$$

The commutative diagram (2.7) then yields a pairing

$$h_r : E_r^{s,t} \otimes F_r^{s',t'} \longrightarrow G_r^{s+s',t+t'}, \quad 2 \leq r \leq \infty$$

which on the E_2 level is $(-1)^{s't}$ times the cup product pairing

$$H^s(X_\infty;H^t(CW;M)) \otimes H^{s'}(X_\infty;H^{t'}(CW,W;M')) \xrightarrow{\cup} H^{s+s'}(X_\infty;H^{t+t'}(CW;W;M''))$$

and which on the E_∞ level is compatible with respect to the cup product pairing

$$H^*(X_\infty;M) \otimes H^*(X_\infty,X_1;M') \xrightarrow{\cup} H^*(X_\infty,X_1;M'')$$

But all three spectral sequences collapse, that is

$$H^s(X_\infty;M) \cong E_\infty^{s,0} \cong E_2^{s,0};\quad H^{s'+n}(X_\infty,X_1;M') \cong F_\infty^{s',n} \cong F_2^{s',n};$$

$$H^{s''+n}(X_\infty,X_1;M'') \cong G_\infty^{s'',n} \cong G_2^{s'',n}$$

and therefore we have the commutative diagram

$$\begin{array}{ccc} H^s(X_\infty;M) \otimes H^n(X_\infty,X_1;M') & \xrightarrow{\cup} & H^{s+n}(X_\infty,X_1;M'') \\ \wr\| & & \wr\| \\ E_2^{s,0} \otimes F_2^{0,n} & \xrightarrow{\quad\cup\quad} & G_2^{s,n} \end{array}$$

Here the bottom row is just

$$H^s(X_\infty;H^0(CW,M)) \otimes H^0(X_\infty;H^n(CW,W;M')) \xrightarrow{\cup} H^s(X_\infty;H^n(CW,W;M''))$$

But there are π module isomorphisms $H^0(CW,M) \cong M$, $H^n(CW,W;M') \cong \mathrm{Hom}(A,M')$, $H^n(CW,W;M'') \cong \mathrm{Hom}(A,M'')$ and so this becomes

$$H^s(X_\infty;M) \otimes H^0(X_\infty;\mathrm{Hom}(A,M')) \xrightarrow{\cup} H^s(X_\infty;\mathrm{Hom}(A,M''))$$

where the coefficient pairing is now given by

$$M \otimes \mathrm{Hom}(A,M') \longrightarrow \mathrm{Hom}(A,M''),\quad m \otimes f \longrightarrow (a \longrightarrow \phi(m \otimes f(a)))$$

For the pairing $\phi : M \otimes M' \longrightarrow M''$ we may choose $\mathrm{Hom}(A,C) \otimes A \longrightarrow C$, $f \otimes a \longrightarrow f(a)$. Thus the following diagram is commutative

$$\begin{array}{ccc} H^s(X_\infty;\mathrm{Hom}(A,C)) \otimes H^n(X_\infty,X_1;A) & \xrightarrow{\cup} & H^{s+n}(X_\infty,X_1;C) \\ \wr\| & & \wr\| \\ H^s(X_\infty;\mathrm{Hom}(A,C)) \otimes H^0(X_\infty;\mathrm{Hom}(A,A)) & \xrightarrow{\cup} & H^s(X_\infty;\mathrm{Hom}(A,C)) \end{array}$$

where now the coefficient pairing $\mathrm{Hom}(A,C) \otimes \mathrm{Hom}(A,A) \longrightarrow \mathrm{Hom}(A,C)$ has become composition. But $H^0(X_\infty;\mathrm{Hom}(A,A))\cong \mathrm{Hom}_\pi(A,A)$ and it is clear that

$$H^s(X_\infty;\mathrm{Hom}(A,C))\xrightarrow[\cong]{\cdot\cup 1_A} H^s(X_\infty;\mathrm{Hom}(A,C)) \text{ for all } s$$

Therefore we have proved

<u>(2.8) Theorem:</u> Assuming the vanishing hypothesis we have

$$\cdot\cup T: H^s(X_\infty;\mathrm{Hom}(A,C))\longrightarrow H^{s+n}(X_\infty,X_1;C)$$

is an isomorphism for all s, where $T\varepsilon H^n(X_\infty,X_1;A)$ is the cohomology class corresponding to $1_A:A\longrightarrow A$ under the isomorphism $H^n(X_\infty,X_1;A)\cong \mathrm{Hom}_\pi(A,A)$.

If $i:X_\infty \rightarrowtail (X_\infty,X_1)$ is the inclusion and $\Delta\varepsilon H^n(X_\infty;A)$ is the cohomology class $i^*(T)$ then we have the commutative diagram

$$\begin{array}{ccc} & \xrightarrow[\cong]{\cdot\cup T} & H^{s+n}(X_\infty,X_1;C) \\ H^s(X_\infty;\mathrm{Hom}(A,C)) & & \Big\downarrow i^* \\ & \xrightarrow{\cdot\cup\Delta} & H^{s+n}(X_\infty;C) \end{array}$$

But X_1 is n-1 dimensional and so $i^*:H^{s+n}(X_\infty,X_1;C)\longrightarrow H^{s+n}(X_\infty;C)$ is an isomorphism for s>0 and an epimorphism for s=0. Therefore we have the reduction isomorphism

$$H^s(\pi;\mathrm{Hom}(A,C))\xrightarrow[\cong]{\cdot\cup\Delta} H^{s+n}(\pi;C) \text{ for } s>0$$

Moreover, it now follows that the epimorphism $\mathrm{Hom}_\pi(A,C)\longrightarrow H^n(\pi;C)$ in (2.6) is cupping with Δ. Therefore we have proved

<u>(2.9) Theorem:</u> Suppose W satisfies the vanishing hypothesis. If A is the π module $H_n(CW,W)$ then there exists a cohomology class $\Delta\varepsilon H^n(\pi;A)$ such that for any π module C

$$H^s(\pi;\mathrm{Hom}(A,C))\xrightarrow[\cong]{\cdot\cup\Delta} H^{s+n}(\pi;C) \text{ for } s>0$$

For s=0 we have the exact sequence

$$\mathrm{Hom}_\pi(C_{n-1}(W),C)\longrightarrow \mathrm{Hom}_\pi(A,C)\xrightarrow{\cdot\cup\Delta} H^n(\pi;C)\longrightarrow 0$$

To prove that the homological isomorphism (2.2) is given by cap product with $T\varepsilon H^n(X_\infty,X_1;A)$ we must study a cap product pairing of spectral sequences. Thus we associate to (2.7) the three spectral sequences

$$E^2_{s,t}\cong H_s(X_\infty;H_t(CW,W;B))\Longrightarrow H_{s+t}(X_\infty,X_1;B)$$

$$F_2^{s',t'}\cong H^{s'}(X_\infty;H^{t'}(CW,W;A))\Longrightarrow H^{s'+t'}(X_\infty,X_1;A)$$

$$G^2_{s'',t''} \cong H_{s''}(X_\infty;H_{t''}(CW;B\otimes A)) \Longrightarrow H_{s''+t''}(X_\infty;B\otimes A)$$

By using the cap product pairings on the base and fibre we obtain a spectral sequence pairing

$$h_r: E^r_{s,t} \otimes F_r^{s',t'} \longrightarrow G^r_{s-s',t-t'},\ 2\leq r\leq\infty$$

which on the E^2 level is the cap product pairing

$$H_s(X_\infty;H_t(CW,W;B))\otimes H^{s'}(X_\infty;H^{t'}(CW,W;A)) \xrightarrow{(-1)^{s't}\cap} H_{s-s'}(X_\infty;H_{t-t'}(CW;B\otimes A))$$

and which on the E^∞ level is compatible with the cap product pairing

$$H_*(X_\infty,X_1;B)\otimes H^*(X_\infty,X_1;A) \xrightarrow{\cap} H_*(X_\infty;B\otimes A)$$

But again all three spectral sequences collapse because of the vanishing hypothesis and so we have the commutative diagram

$$\begin{array}{ccc} H_{s+n}(X_\infty,X_1;B)\otimes H^n(X_\infty,X_1;A) & \xrightarrow{\cap} & H_s(X_\infty;B\otimes A) \\ \wr\Vert & & \wr\Vert \\ E^2_{s,n}\otimes F_2^{0,n} & \xrightarrow{\cap} & G^2_{s,0} \end{array}$$

The bottom line of this diagram is

$$H_s(X_\infty;H_n(CW,W;B))\otimes H^0(X_\infty;H^n(CW,W;A)) \xrightarrow{\cap} H_s(X_\infty;H_0(CW;B\otimes A))$$

But $H_n(CW,W;B)\cong B\otimes A$, $H^n(CW,W;A)\cong \mathrm{Hom}(A,A)$ and $H_0(CW;B\otimes A)\cong B\otimes A$ so this pairing becomes

$$H_s(X_\infty;B\otimes A)\otimes \mathrm{Hom}_\pi(A,A) \xrightarrow{\cap} H_s(X_\infty;B\otimes A)$$

Since capping with $1_A:A\longrightarrow A$ clearly gives the identity isomorphism $H_s(X_\infty;B\otimes A)\longrightarrow H_s(X_\infty;B\otimes A)$ we have proved

<u>(2.10) Theorem:</u> Assuming the vanishing hypothesis we have

$$\cdot\cap T: H_{s+n}(X_\infty,X_1;B) \longrightarrow H_s(X_\infty;B\otimes A)$$

is an isomorphism for all s, where $T\in H^n(X_\infty,X_1;A)$ is the cohomology class corresponding to $1_A:A\longrightarrow A$ under the isomorphism $H^n(X_\infty,X_1;A)\cong \mathrm{Hom}_\pi(A,A)$.

Now consider the commutative diagram

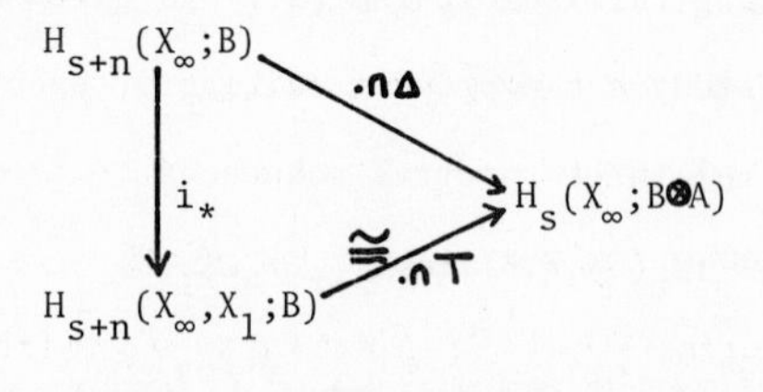

Because $i_*:H_{s+n}(X_\infty;B)\longrightarrow H_{s+n}(X_\infty X_1;B)$ is an isomorphism for s>0 and a monomorphism for s=0 we have

$$H_{s+n}(\pi;B)\xrightarrow[\cong]{\cdot\cap\Delta}H_s(\pi;B\otimes A)\quad\text{for } s>0$$

Finally the monomorphism $H_n(\pi;B)\longrightarrow B\otimes_\pi A$ in (2.5) is now seen to be capping with Δ.

(2.11) Theorem: Suppose W satisfies the vanishing hypothesis and $A=H_n(CW,W)$. Then there exists $\Delta\varepsilon H^n(\pi;A)$ such that

$$H_{s+n}(\pi;B)\xrightarrow[\cong]{\cdot\cap\Delta}H_s(\pi;B\otimes A)\quad\text{for } s>0$$

For s=0 we have the exact sequence

$$0\longrightarrow H_n(\pi;B)\xrightarrow{\cdot\cap\Delta}B\otimes_\pi A\longrightarrow B\otimes_\pi C_{n-1}(W)$$

(2.12) Remark: Many of the theorems in this section are true under assumptions weaker than those of the vanishing hypothesis. For example, theorems (2.2), (2.8) and (2.10) remain valid if we only assume that there exists an integer n such that

$$H_t(CW,W)=0 \text{ for } t\neq n \text{ and } A=H_n(CW,W) \text{ is torsion free.}$$

§3 Applications:

As our first application consider the case $W=\pi$ with π acting on itself by left multiplication. The vanishing hypothesis holds for n=1 and the chain complex $C_*(CW,W)$ is

$$0\rightarrow Z[\pi]\xrightarrow{\varepsilon}Z\longrightarrow 0$$

where $Z[\pi]$ is the integral group ring of π and ε is the usual augmentation. Thus $H_1(CW,W)$ is the augmentation ideal $I[\pi]$ and we have the following corollary of (2.9), (2.11):

(3.1) Corollary: There exists a cohomology class $\Delta\varepsilon H^1(\pi;I[\pi])$ such that for all π modules B, C we have the isomorphisms for s>0

$$H^s(\pi;Hom(I[\pi],C))\xrightarrow{\cdot\cup\Delta}H^{s+1}(\pi;C);\ H_{s+1}(\pi;B)\xrightarrow{\cdot\cap\Delta}H_s(\pi;B\otimes I[\pi])$$

For s=0 we have the exact sequences

$$Hom_\pi(Z[\pi],C)\longrightarrow Hom_\pi(I[\pi],C)\xrightarrow{\cdot\cup\Delta}H^1(\pi;C)\longrightarrow 0$$

$$0\longrightarrow H_1(\pi;B)\xrightarrow{\cdot\cap\Delta}B\otimes_\pi I[\pi]\longrightarrow B\otimes_\pi Z[\pi]$$

For our second application suppose we are given a free presentation $1\rightarrow R\rightarrow F\rightarrow\pi\rightarrow 1$ of the group π. Then, by the theory of covering spaces, we can construct a regular covering $W\longrightarrow X_1$ with deck transformation group π and such that W, X_1 are Eilenberg-MacLane spaces of the respective types K(R,1),

$K(F,1)$. The vanishing hypothesis holds for $n=2$ and $H_2(CW,W)$ is the standard π module $R_{ab}=R/[R,R]$. Therefore we have

(3.2) Corollary: There exists a cohomology class $\Delta \varepsilon H^2(\pi;R_{ab})$ such that for all π modules B,C we have for s>0 the following isomorphisms

$$H^s(\pi;\mathrm{Hom}(R_{ab},C)) \xrightarrow[\cong]{\cdot\cup\Delta} H^{s+2}(\pi;C),\ H_{s+2}(\pi;B) \xrightarrow[\cong]{\cdot\cap\Delta} H_s(\pi;B\otimes R_{ab})$$

The two corollaries above are the familiar reduction theorems of homological algebra. For a similar application suppose X is a $(\pi,n-1)$ complex, $n\geq 3$; that is X is a connected n-1 dimensional complex such that $\pi_1(X)\cong\pi$ and $\pi_i(X)=0$ for $1<i<n-1$. For W we take the universal covering space $\tilde{X}$. Then the vanishing hypothesis holds and we have

$$A=H_n(CW,W)\cong H_{n-1}(W)\cong \pi_{n-1}(W)\cong \pi_{n-1}(X)$$

Thus we have

(3.3) Corollary: Suppose X is a $(\pi,n-1)$ complex, $n\geq 3$. Then there exists a cohomology class $\Delta\varepsilon H^n(\pi;\pi_{n-1}(X))$ such that for all π modules B,C we have the isomorphisms

$$\cdot\cup\Delta : H^s(\pi;\mathrm{Hom}(\pi_{n-1}(X),C)) \xrightarrow{\cong} H^{s+n}(\pi;C) \quad \text{for} \quad s>0$$

$$\cdot\cap\Delta : H_{s+n}(\pi;B) \xrightarrow{\cong} H_s(\pi;B\otimes\pi_{n-1}(X)) \ \text{for}\ s>0$$

This corollary includes some of the results of [2]. For our last application suppose given a group extension $1\longrightarrow N\xrightarrow{\mu}G\xrightarrow{\varepsilon}\pi\longrightarrow 1$, where we assume

(3.4)
N has trivial cohomological dimension 1
G has finite cohomological dimension k

Thus $H_i(N)=0$ for all $i\geq 2$ and $H_1(N)$ is a non-trivial torsion free abelian group. As examples of such groups N we have

(3.5)
(i) free groups;
(ii) the group of a knot tamely embedded in S^3 [4];
(iii) any one relator group N for which $H_1(N)$ is non-trivial free abelian and such that the relation is not a product of commutators [3];
(iv) free products of the above.

Now realize the extension by a regular covering $W\longrightarrow X_1$ with deck transformation group π. Thus W is a $K(N,1)$, X_1 is a $K(G,1)$, X_∞ is a $K(\pi,1)$, and the inclusion $X_1 \rightarrowtail X_\infty$ is induced by $\varepsilon: G \to \pi$. According to our remark (2.12) the main theorems of §2 continue to hold. In particular there exists a cohomology class $T \in H^2(X_\infty, X_1; N_{ab})$ such that for all π modules B, C we have

$$\cdot \cup T : H^s(X_\infty; \mathrm{Hom}(N_{ab}, C)) \to H^{s+2}(X_\infty, X_1; C) \text{ is an isomorphism for all } s$$

$$\cdot \cap T : H_{s+2}(X_\infty, X_1; B) \to H_s(X_\infty; B \otimes N_{ab}) \text{ is an isomorphism for all } s.$$

If $i: X_\infty \subset (X_\infty, X_1)$ is the inclusion we have $i^*: H^t(X_\infty, X_1; C) \to H^t(X_\infty; C)$ is an isomorphism for $t > k+1$ and an epimorphism for $t = k+1$. Similarly $i_*: H_t(X_\infty; B) \to H_t(X_\infty, X_1; B)$ is an isomorphism for $t > k+1$ and a monomorphism for $t = k+1$. Putting $\Delta = i^*(T) \in H^2(\pi; N_{ab})$ then gives

<u>(3.6) Corollary:</u> Assuming the hypothesis (3.4) we have for all π modules B, C

$\cdot \cup \Delta : H^s(\pi; \mathrm{Hom}(N_{ab}, C)) \to H^{s+2}(\pi; C)$ is an isomorphism for $s > k-1$ and an epimorphism for $s = k-1$;

$\cdot \cap \Delta : H_{s+2}(\pi; B) \to H_s(\pi; B \otimes N_{ab})$ is an isomorphism for $s > k-1$ and a monomorphism for $s = k-1$.

These morphisms also satisfy a certain functoriality condition in the group π. If π' is a subgroup of π we have the induced extension $1 \to N \to G' \to \pi' \to 1$, $G' = \varepsilon^{-1}(\pi')$. Then G' has cohomological dimension $\leq k$ and so we have (3.6) holding in this situation. Moreover, the class $\Delta' \in H^2(\pi'; N_{ab})$ is $j^*(\Delta)$, where $j: \pi' \subset \pi$ is the inclusion, and the following diagrams commute

$$\begin{array}{ccc} H^s(\pi; \mathrm{Hom}(N_{ab}, C)) & \xrightarrow{\cdot \cup \Delta} & H^{s+2}(\pi; C) \\ \downarrow j^* & & \downarrow j^* \\ H^s(\pi'; \mathrm{Hom}(N_{ab}, C)) & \xrightarrow{\cdot \cup \Delta'} & H^{s+2}(\pi'; C) \end{array} \qquad \begin{array}{ccc} H_{s+2}(\pi; B) & \xrightarrow{\cdot \cap \Delta} & H_s(\pi; B \otimes N_{ab}) \\ \uparrow j^* & & \uparrow j^* \\ H_{s+2}(\pi'; B) & \xrightarrow{\cdot \cap \Delta'} & H_s(\pi'; B \otimes N_{ab}) \end{array}$$

As a special case of (3.6) we have

<u>(3.7) Corollary:</u> Suppose $1 \to N \to G \to \pi \to 1$ is an exact sequence of groups, where G has finite cohomological dimension k, $N_{ab} \cong Z$ as trivial π modules, and $H_i(N) = 0$ for all $i \geq 2$. Then there exists $\Delta \in H^2(\pi)$ such that for all π modules B, C we have

$\cdot\cup\Delta: H^s(\pi;C) \longrightarrow H^{s+2}(\pi;C)$ is an isomorphism

for $s>k-1$ and an epimorphism for $s=k-1$;

$\cdot\cap\Delta: H_{s+2}(\pi;B) \longrightarrow H_s(\pi;B)$ is an isomorphism

for $s>k-1$ and a monomorphism for $s=k-1$.

In addition, these periodicity morphisms are functorial with respect to subgroups of π.

This corollary prompts us to make the following definition.

(3.8) Definition: A group π(not necessarily finite) is said to be eventually periodic with period m if there exists a cohomology class $\Delta\varepsilon H^m(\pi)$ and an integer 1 such that

(i) for all π modules C the homomorphism $\cdot\cup\Delta: H^s(\pi;C) \longrightarrow H^{s+m}(\pi;C)$ is an isomorphism for $s\geq 1$;

(ii) if π' is any subgroup of π, C' is any π' module, and $\Delta'=\Delta/\pi'$ then $\cdot\cup\Delta': H^s(\pi';C') \longrightarrow H^{s+m}(\pi';C')$ is an isomorphism for all $s\geq 1$.

Corollary (3.7) gives an example of a group π which is eventually periodic with period 2. If in (3.7) we only assume that $H_1(N) \cong Z$ as abelian groups, and not necessarily as π modules, then it will follow that π is eventually periodic with period 4. To see this we apply (3.6) twice.

(3.9) Corollary: Assuming the hypothesis of (3.7) it follows that all finite subgroups of π are cyclic.

Proof: For any positive integer d the cohomology of $Z_d \oplus Z_d$ is not eventually periodic, and so all finite abelian subgroups of π are cyclic. In other words, all finite subgroups of π are periodic, and since the period must be 2 it follows that all finite subgroups are cyclic (see chapter 12 of [1]).

(3.10) Example: Suppose $a_1,\cdots,a_n$ are non-zero integers. Then the group presented by $G=\{x_1,\cdots,x_n \mid x_1^{a_1}=\cdots=x_n^{a_n}\}$ has cohomological dimension ≤ 2 and its centre is the infinite cyclic subgroup N generated by $x_1^{a_1}$. Thus (3.7) applies and we conclude that the free product $Z_{a_1} * \cdots * Z_{a_n}$ of cyclic groups is eventually periodic with period 2.

References

1. Cartan and Eilenberg, Homological Algebra, Princeton University Press, 1956.

2. Dyer, M., On the Second Homotopy Module of Two Dimensional CW Complexes, Proc. Amer. Math. Soc. 55(1976)400-404.

3. Gruenberg, Cohomological Topics in Group Theory, Lecture Notes in Mathematics 143, Springer-Verlag 1970.

4. Papakyriakopoulos, On Dehn's Lemma and the Asphericity of Knots, Ann. of Math. 66(1957)1-26.

5. Sjerve, Group Actions and Generalized Periodicity, Math. Z. 149, 1-11(1976).

6. Sjerve, The Thom Space Periodicity of Classifying Spaces, Proc. Amer. Math. Soc. 65(1977)165-170.

7. Swan, Periodic Resolutions for Finite Groups, Ann. of Math. 72(1960)267-291.

ADDRESSES OF CONTRIBUTORS

Professor Peter Booth
Memorial University of Newfoundland
Department of Mathematics,
Statistics and Computer Science
St. John's, Newfoundland
A1B 3X7 Canada

Professor Albrecht Dold
Universität Heidelberg
Mathematisches Institut
Heidelberg, West Germany

Professor Roy Douglas
University of British Columbia
Department of Mathematics
Vancouver, British Columbia
V6T 1W5 Canada

Professor Henry Glover
Ohio State University
Department of Mathematics
Columbus, Ohio
43210 U.S.A.

Professor Philip Heath
Memorial University of Newfoundland
Department of Mathematics,
Statistics and Computer Science
St. John's, Newfoundland
A1B 3X7 Canada

Professor Peter Hilton
Battelle Seattle Research Center
4000 N.E. 41st
Seattle, Washington
98105 U.S.A.

Professor Bill Homer
Ohio State University
Department of Mathematics
Columbus, Ohio
43210 U.S.A.

Professor Richard Kane
University of Alberta
Department of Mathematics
Edmonton, Alberta
T6G 2E1 Canada

Professor Arunas Liulevicius
University of Chicago
Department of Mathematics
Chicago, Illinois
60637 U.S.A.

Professor Guido Mislin
Lehrstuhl für Mathematik
ETH
Zürich, Switzerland

Professor Renzo Piccinini
Memorial University of Newfoundland
Department of Mathematics,
Statistics and Computer Science
St. John's, Newfoundland
A1B 3X7

Professor Joseph Roitberg
City University of New York
Hunter College
Department of Mathematics
New York, N. Y.
10021 U.S.A.

Professor Jack Segal
University of Washington
Department of Mathematics
Seattle, Washington
98195 U.S.A.

Professor François Sigrist
Université de Neuchâtel
Institut de Mathématiques
Chantemerle 20
Neuchâtel, CH 2000, Switzerland

Professor Denis Sjerve
University of British Columbia
Department of Mathematics
Vancouver, British Columbia
V6T 1W5 Canada

Professor Victor Snaith
University of Western Ontario
Department of Mathematics
London, Ontario
N6A 5B9 Canada

Professor James Stasheff
University of North Carolina
at Chapel Hill
Department of Mathematics
Chapel Hill, North Carolina
27514 U.S.A.

Professor Richard Steiner
University of Western Ontario
Department of Mathematics
London, Ontario
N6A 5B9 Canada

Professor Ueli Suter
Université de Neuchâtel
Institut des Mathématiques
Chantemerle 20
Neuchâtel, CH 2000, Switzerland

Professor Martin Tangora
University of Illinois
at Chicago Circle
Department of Mathematics
Chicago, Illinois
60680 U.S.A.

PARTICIPANTS

J.Allard
D.Bass
R.Body
P.Booth
E.Campbell
F.Cathey
A.Dold
R.Douglas
S.Feder
R.Gentle
H.Glover
P.Heath
P.Hilton
P.Hoffman
W.Holzmann
S.Jajodia
R.Jardine
R.Kane
A.Liulevicius
W.MacIlquham
D.Mauro
J.McCleary
G.Mislin
C.Morgan
S.Nanda
D.Pengelley
R.Piccinini
L.Renner
J.Segal
L.Siebenmann
F.Sigrist
D.Sjerve
V.Snaith
J.Stasheff
P.Stone
U.Suter
M.Tangora
J.Tillotson
J.Timourian
J.Verster
C.Watkiss
J.Wittaker